ALL DONE WITH MIRRORS

All Done With Mirrors

A look back to a time of making telescopes

A look ahead to an advance in watching the sky

Observations, photographs, and illustrations

by

Roy Kaelin

E-BookTime, LLC
Montgomery, Alabama

All Done With Mirrors

Copyright © 2012 by Roy Kaelin

Cover image: The light of the waxing crescent Moon as seen through the optics of the author's Mersenne telescope. Photo by Roy Kaelin.

All rights reserved. No part of this book may be reproduced or transmitted in any form or by any means, electronic or mechanical, including photocopying, recording, or by any information storage and retrieval system, without permission in writing from the copyright owner.

Library of Congress Control Number: 2012900127

ISBN: 978-1-60862-362-4

First Edition
Published January 2012
E-BookTime, LLC
6598 Pumpkin Road
Montgomery, AL 36108
www.e-booktime.com

Contents

Prologue 7

Foreword 9

Chapter 1
Once Upon A Star

Why, oh why, build your own telescope? 13
And is that the telescope you really want? 17

Chapter 2
Barrels of Fun

Round and round we go! 24
Does the ideal telescope exist? 28

Chapter 3
Grinding Passion

A worthwhile goal, your first telescope 37
A worthwhile design, the Mersenne-Nasmyth 45

Chapter 4
Smoothing the Curve

Mirror-making becomes rewarding 56
The Mersenne-Nasmyth offers a rewarding
    alternative 57

Chapter 5
Keep Focused

Are we there yet? 60
How is a Mersenne scope any better? 63

Chapter 6
Go Figure

That wasn't there a moment ago! 67
What makes a Mersenne design work? 72

Chapter 7
The Victory Lap

Then comes that magic moment 80
How does this Mersenne design really work? 85

Chapter 8
One Happy Time

Gilded memories 95
Make new memories for yourself 96

Epilogue 103

References 106

Prologue

By 1957, the Adler Planetarium and Astronomical Museum, as it was then known, had opened the Amateur Telescope Makers Optical Shop to its members and the visiting public. A living exhibit tucked to one side of the planetarium's former mid-level, the Optical Shop provided inquisitive onlookers with the chance to watch amateur astronomers grind and polish glass disks into mirrors suitable for use in homebuilt telescopes to view the heavens.

From that humble start and for the next 40 years, families and individuals who worked 'round the barrels in the Optical Shop discovered the sheer enjoyment and personal satisfaction of crafting their very own optics with which to observe the night sky. Fond memories certainly abound of those happy times, and this little book is merely a reflection of the fine instructors and the dedicated amateurs who found both solace and camaraderie in the art of mirror-making. It is also a paean to the practical inspiration and actual accomplishment to which crafting one's own mirror and telescope can lead.

To my fellow amateur telescope-makers, this small endeavor is dedicated to your unstinting effort in telescope-making to hone your technique, advance your art, and craft a workable instrument with which to discover the night sky. I hope this little book offers a fair perspective of an avocation that captured our heart and mind when we first set eyes on the skies. I also hope that the distinctive telescope designs presented herein, inspired from that love affair with the heavens, help to set many more eyes on the night skies as well.

Foreword

Poignant quotations are often cited at transcendent moments, as in war or in peace, as in times of greatness or in times of duress, as in the celebration of life or in the remembrance of death. For those other moments, individuals often find the appropriate words necessary to convey the importance of what may seem everyday, but no less mundane.

For the makers of telescopes and observers of stars, words may not always convey the poignancy with which they experience their avocation. Some may be reticent, quietly content to keep to themselves the personal moments of building a telescope or of scanning the night skies. More often than not, others are more candid and are willing to relate every tedious detail describing the manufacture of a homebuilt instrument or of last evening's observing session. Somewhere between these two extremes, appropriate words surely can be found to recount the experiences of those who were privileged to craft a mirror of their own making at the once and valued Amateur Telescope Makers Optical Shop. This little book offers a sampling of those memorable experiences.

Coincident with the launch of the Soviet Union's first Sputnik satellite and the commencement of the International Geophysical Year, the Amateur Telescope Makers Optical Shop was well underway at the Adler

Planetarium and Astronomical Museum, as it was then known. On a lower level of the Planetarium, visitors could watch as hobbyists would grind, polish, test, and transform a small disk of glass into a shiny telescope mirror of fairly exacting standards. In a procedure that often appeared medieval in its execution, the amateur telescope-maker, the ATM, enjoyed learning those scientific principles, more often the practical insights of an optician's art, but still based on the physics of optics.

In later decades, the Optical Shop attracted an eager following, as amateur telescope-makers often returned, some with their own children in tow, to experience again the craft that allowed them to open their first close-up window on the night sky. Under the tutelage of a variety of instructors over the years, the ATM became adept at a mechanic art that allowed him to fashion a curved, optically correct glass often more accurate than was commercially available at the time. And, with a reputation for precise mirrors, guided to completion by able instructors, such practice became a hallmark of the Optical Shop's dependable accuracy.

But this book is not about the Optical Shop or the Adler Planetarium. This little book does portray both at several moments in time; however, my intent is not to offer an historical point of view of their founding or even an elementary retelling of their origin. Rather, their inclusion is just a brief retrospective of a time and a place that set the stage for greater learning later. Certainly, the methods of craftsmanship gleaned from the Optical Shop were valuable, but the insights gained from their consistent practice led to a better ability to advance the art of telescope-making. What was taught at the barrel in that shop formed the basis for breakthroughs at a later date; and those breakthroughs were both unexpected and welcome.

Foreword

What this little book offers is a perspective on the utility of practicing to think, on the serendipity of unexpected insight, on the value of camaraderie. That the Adler Planetarium was a suitable stage and the Optical Shop an appropriate set on that stage ultimately doesn't matter here. They served a useful purpose certainly, but a separate stage or another set in a different city could have just as easily provided the direction and impetus for lessons learned and insights gained. And that probably did happen for a number of you reading this, who sought to advance an interest, pursue a hobby, or satisfy an avocation by becoming an ATM. No doubt you have your own memorable moments if you were privileged to spend time at a planetarium or in the company of sincere individuals with a shared interest in telescope-making.

In this little book, of course I highlight certain memorable moments of my own time around the barrel in the Optical Shop. That's only normal and natural to do so, especially if I'm describing a scene of a fine time spent among valued friends and earnest acquaintances. But those moments are largely incidental to the simple aim of this little book.

My aim is this: To show that a solid development often sprouts from a humble seed. That development involves the growth and advance of distinct telescope design that got its start as an inkling of a grand, glassy instrument, as an offshoot of trial-and-error attempts, as a way to improve the observation of the night sky with comfortable access to a large aperture.

That's what this little book reveals. It is about amateur telescope-making at its most inquisitive. It is about crafting a traditional design, but advancing to two more novel configurations. It is about making two distinct optical surfaces, the first one out of curiosity, the newer one, more like an oddity. It is also a tale of two mirrors: one that was my first; the newer one, my best.

I also make known highlights of the novel design's development and construction almost as if we're building a reflector itself. Though this book is meant for the curious hobbyist, it has detail for the serious enthusiast. So, in each short chapter, I offer bits of telescope-making lore, personal experiences in crafting a workable reflecting telescope, excerpts from past articles I've prepared on telescope designs, and references for further reading on mirror-making. By the time we're through, you'll know how I built my first scope, how to avoid a few of the pitfalls in making your own scope, and, how to deepen your exploration of the night sky with a distinctive telescope that is, to use a phrase, all done with mirrors.

Roy Kaelin
August 9, 2011

Chapter 1

Once Upon A Star

Why, oh why, build your own telescope?

What a genuine joy it was at that young age, to take my first step toward securing my very own telescope. What I sought to achieve was no less than to master the heavens. It was with a telescope of my very own that I sought to achieve that goal.

I was fourteen years old at the time in that spring of 1969 and had just emerged into the afternoon light from the dim, cavernous exhibit halls at the Adler Planetarium. I was so pleased and proud that I finally was on my way! But on my way to what? I had simply purchased a small cardboard box for $6.50 at the planetarium's gift counter and now I wanted to see what was in it. Inside the box was a mirror-making kit, or so it said on the outside of the box. A kit for making a six-inch telescope. Not a telescope six inches high, of course, but for that telescope a glass eye six inches wide. I headed to the bronze and glass entrance and pushed through the first set of heavy doors.

Sunlight streamed through the green leaded glass of the building's outer doors and scattered rainbows high across the granite walls within the front foyer. For the moment I noticed that the great slabs of polished stone lining that foyer had matching faces, as if the front half of the wide walls leading to the outside were a mirror image of the back half leading to the inside. Just like two great pieces of matching glass that become polished surfaces. Just like what an amateur telescope-maker seeks to achieve.

I steadied the trim box under my arm and pushed through the second set of bronze doors. Down the wide, smooth, stone steps I bounded, careful not to slip, and climbed atop a boxy granite slab to one side of the Planetarium's grand outdoor staircase. Eager to lay eyes on my prize, I still hesitated for a moment and didn't open the little box. I knew I had yet to get to the train and to get the box home; I certainly didn't want to lose or break anything on the way. I savored opening my treasure till later.

As I headed to catch the Illinois Central commuter train at the nearby station, known then as 12th Street & Roosevelt Road, I shot a glance back at the Planetarium down the main boulevard, known then as Achsah Bond Drive. (See Photo 1 for an outside picture of Adler, circa 1970.) That domed edifice appeared a sweet sight, a hump of chocolate-brown against the brilliant blue of that wonderful springtime afternoon. Far sweeter, I recall, would be the day I would see my own telescope cut its own pleasing profile against the sky.

Photo 1

The Adler Planetarium opened in 1930 on Northerly Island in Lake Michigan. When this picture was taken in 1970, on a lower level was located the Amateur Telescope Makers Optical Shop, where, at the time, the author worked on his first mirror.

* * *

Like so many starry-eyed kids my age who gazed at the night sky, I so wanted a telescope. There were fancy lens telescopes on display in a few magazines, including issues of *Sky and Telescope*. A popular brand of refractor at the time was Unitron, a Japanese model. Complete with setting circles and a glass Sun filter – green in color, deadly to use – that screwed into its eyepieces, it seemed the envy of almost any amateur. However, it didn't come at an amateur price at the time.

I had used almost any optical aid I could find, including a toy refractor I'd received one autumn as a birthday present from a school chum when I was eleven. That was my first telescope: two tin tubes, one nested within the other, wrapped with a black crinkle vinyl finish, holding a set of small glass lenses, gimbaled by a screw on a tiny metal tripod. At first, I was enthused to have it, and recall setting it on a concrete patio bench in the backyard at my parents' house. That toy telescope aimed skyward, I anticipated grand cosmic vistas to open up before my eyes. However, my first look through that little toy revealed no grand cosmic vistas. It showed barely anything at all, just a shimmering blob where the star chart said the star Vega should be.

While my first view through a toy telescope began once upon a star, and not a very good view at that, I wasn't discouraged. It was then and there that I realized what the astronomy books from the library uptown meant about how a telescope gathers light and magnifies it. This toy telescope did neither very well. I knew I needed more.

My mother owned a pair of opera glasses, which folded neatly into a black leatherette box rimmed in shiny chrome. Borrowing these, I saw some sky things a little better, though a neat cluster of stars, known as the Pleiades, seemed only slightly larger when I witnessed them rising above the trees in the cold air of a Midwestern winter. I could figure out that the magnification of these little glasses let me see things at least three times closer than they appeared. Of course I needed more.

There were the reflectors, the mirrored telescopes, one of the first of which was built by Sir Isaac Newton. As interested as any schoolboy might be in history and astronomy, I was certainly familiar with the historic Newton, including the story of his inspiration from a falling apple. I felt drawn to his design of that first reflecting telescope, as it embodied all of the neat features that made it both compact and practical. To that end, I used to draw how one might appear were I to build it, depicting my own cutaway diagrams of the Newtonian design to see how it worked, and penciled for myself the traces of light rays reflected from surface to mirrored surface.

The Adler Planetarium once had on display a replica of Newton's own scope. The small model seemed so unassuming, so humble, so puny compared to the larger scopes advertised in *Sky and Telescope* magazine and elsewhere. That little replica's clear-cut, neatly done, home-brew appearance is what made it so appealing. That's what I knew I needed. So, a reflecting telescope seemed the best, and only, route for me as a homebuilt instrument.

My attention to astronomy snapped to at an early age, likely as a result of visits to the Adler Planetarium with my parents. It seemed a Chicagoland ritual, but many still visit the renowned landmark. As one wag who worked there once remarked, and as others have since, there are at least three times in a person's life when one visits the Planetarium: as a child, as parents with children, and as grandparents with the grandkids. I suppose we visit as often we like, or as often as we think something there is worthwhile to see, but perhaps not as often as the folks who work there would like. I understand that, too, because I once worked there.

* * *

And is that the telescope you really want?

I met Kenneth Wolf at the Optical Shop in late 1970. I never knew him well as I only saw him on certain days when I visited the Planetarium, so I wouldn't know his day-to-day opinion of things or of people; I only sought him out because at the time he was the man to see about building telescopes and making mirrors. Ken Wolf was the director of the Optical Shop at the Adler Planetarium.

When I first met him, he was wedged into a time-worn office chair rolled up against a gray steel desk in a tiny, narrow office, almost a hallway it seemed, between the grinding room and the polishing room of the Optical Shop. I had stopped by to inquire about a course in mirror-making and, matter-of-factly, he let me know the shop's usual hours. It was a Saturday afternoon, which I recall was a popular time for amateur telescope-makers to work at the shop.

At the time, I was also a high-school student, participating in the Planetarium's Astro-Science Workshop (ASW), a cooperative venture between the National Science Foundation and selected schools, such as Northwestern University, which held its weekly ASW lectures at the Planetarium on Saturday mornings during the regular school year. Guest lecturers were occasionally invited throughout the regular term to the Planetarium, though the full set of classes culminated with ASW's Christmas Lecture – orchestrated by Miss Letitia Lestina who served as the ASW program's executive secretary – and presented in old Thorne Hall on the Northwestern campus. Such a delight it was at the time to hear renowned astronomers such as Bart Bok and physicists such as Sir John Eccles and Philip Morrison deliver their captivating and erudite lectures. Such an opportunity, too, it was for an eager teen with a promising pursuit of all things astronomical.

On this particular Saturday, the other students and I had listened that morning to a short lecture by Dr. J. Allen Hynek, Chairman of Northwestern's Astronomy Department at the time, in the Planetarium's lecture hall, and, we'd received an assignment to measure galactic spectrum lines in order to compute the velocities at which these galaxies were moving through space.

From our careful measurements and eventual computations we were to determine whether these galaxies were approaching or receding. It was typical of the kind of techniques astronomers employed to measure the extent of the known universe. It was the intent of the assignment to prove to ourselves Edwin Hubble's law regarding expansion of the universe. It was fascinating to have copies of actual astronomical data in hand, but, from my parents' backyard and without a telescope large enough, it was certain I'd never glimpse the distant galaxies listed in the assignment.

That day, as on so many other visits, I'd scouted the back hallway of the Planetarium's uppermost floor and gazed at backlit observatory photographs of distant galaxies and nebulae, pondering their origin, their size, and their distance. Of course I doubted my humble telescope, the one I'd started from a kit, would ever see galaxies in such magnificent array, but I still wondered what I'd eventually see with my own scope. Or with a larger scope one day.

From the lecture hall on the lower level I headed across the terrazzo floor of the exhibit hall to two large glass windows on the other side of the hall. To me, beyond that glass laid undiscovered country and the Optical Shop itself. An array of drums – known as barrels, in telescope-making parlance – dotted the floor behind the two great picture windows. Under the blaze of fluorescent lighting, several barrels were already in service, and those working 'round the barrels certainly seemed intent on their hobby. But even as I was still learning the art of those keen enthusiasts, I grasped at once the need to join them behind the glass and learn more.

At the time, I also felt I needed to join them pretty soon, too, since my own telescope mirror just wasn't what the books said it should be. My own six-inch mirror was becoming a dog; rather, its surface had the undistinguished look of a "dog-biscuit" – at least that's what "Sam Brown" showed in his slim, mirror-making booklet from Edmund Scientific – which was a mirror surface about as bad as one could get in the realm of amateur telescope-making.

It wasn't from lack of diligent work; rather, it was from a lack of the right kind of persistence to recognize that I had a lousy mirror. I just had to take good, slow care to understand more completely what I was doing. So, that's why I sought out Mr. Wolf.

He was cordial though clear, and, after apprising me of the ATM program at the Optical Shop, he signed me up for his mirror-making class right then and there. I returned the following Saturday for my usual ASW class and, afterward, brought him my six-inch glass to inspect. Of course, thanks to a quick, shrewd look at my sorry mirror, he knew at once this one needed work. In his low-key way, he assured me the whole affair wasn't hopeless. Not at all.

I'd already done the opening round of grinding it and smoothing it to the desired curve, but I was impatient to get it polished. At home, I'd set up a makeshift polishing effort, following a book recipe to get it underway, but the outcome so far appeared fitful and uneven. No wonder the mirror had sleeks and patchy polishing that gave it a "dog-biscuit" surface. Of course, without a word, that's what Mr. Wolf already knew.

Rather than let me set foot in the polishing room with its convenient barrels and a view to the outside world of interested onlookers, he sent me into a smaller room, a narrow windowless hallway stocked with its own barrels, that served as the grinding room. Yes, I was going back to grinding! Well, fine grinding at least. Where I might've thought I was ready for final polishing, Ken Wolf knew better. (See Photo 2, pencil illustration of Ken Wolf.)

Photo 2

**Ken Wolf as Director of Optical Shop,
circa 1970**

If I felt discouraged at the time, I quickly got past it, since fine grinding on a barrel – a sand-filled 55-gallon drum actually – at the Optical Shop offered a clearer, well-paced insight toward getting the curve correct, rather than my own set-up on a wobbly bookcase in a second-floor bedroom at home. I'd used my kit's grinding agents, few as they were, because they were available, and perhaps, I concluded, not entirely in the most efficient way.

I'd become accustomed to using the grinding abrasives freely, scooping up and re-using what I could – just as Sam Brown had said – but the use of fine grinding powders could be easily extended, stored, and re-used as a milky slurry. The action of fine grinding, at least in the manner which Mr. Wolf instructed, was delightfully smooth and easygoing. I got more and better work done on the curve with the lighter approach of fine grinding than my attempts at bearing down on a disk of glass, bumping its way through uneven polishing.

By the time I was ready to leave the narrow confines of the grinding room, the surface of my six-inch glass was free of pits on its surface, glossy smooth to the touch, and evenly frosted from the action of micron-fine abrasive. I spent the time needed to get it right, and it paid off, especially since my first mirror was fairly steep for a beginner. Of course that meant more time to get the entire curved surface correct, right out to the edge.

Short-focus mirrors, crafted into a well-made telescope, always cut a pleasing profile, as I saw them. They appeared sturdy, not ungainly. They seemed more stable and easier to balance since the load concentrated at a fulcrum close to the center of a tripod or low to the ground. Yes, they tended to be more useful for distant objects, like star clusters and galaxies, since the mirror's steep curve concentrated most of the gathered light at the center of the field of view, which was all the better to observe faint, point-like objects.

Certainly the long-focus models had their advantages, too, especially if one wanted a grand scope to pursue the planets, since more of the gathered light was spread over the entire field of view to increase the scene's contrast, which was all the better to observe extended objects. Old-time photos of long-focus Newtonians undoubtedly had their appeal, depicting a long eye on the sky, while the observer stood on a tall ladder or scaffold to reach the eyepiece. Or, depending on the style of telescope mounting used, occasionally a long-focus mirror outfitted in an ungainly and long tube, needed a trench in which the observer stood!

I'd always heard and read that the ideal compromise between short-focus observation of deep-sky objects, like galaxies and nebulae, and long-focus views of extended celestial objects, like the Moon and planets, was a six-inch diameter mirror that could fit optically inside the length of a four-foot long tube, which was fairly easy to handle and set up. At the time, I had high-school friends who'd built classic Newtonians of this very size. Those scopes looked good, operated well, and certainly seemed the ideal telescope. For them, but not for me.

From my very first glass disk to the later and larger ones I ground and polished, I chose to craft short-focus mirrors. At the time, I thought that they represented the ideal in telescope design because they looked trim and neat and were fairly easy to transport. They required more work, but my intent always favored portability and ease of set-up, that is, to get the scope out-of-doors and at a nighttime gathering. Even if the telescope's aperture was large, its short-focus length would always ensure its stability and ease of transport. Or so I thought.

Chapter 2

Barrels of Fun

Round and round we go!

My first exposure to mirror-making did not end with the Optical Shop. It was just part of the beginning. I'd already become a subscriber to *Sky and Telescope* magazine the same year I started my first mirror and found some of its monthly features worth perusing for ideas and encouragement, especially the column "Gleanings for ATMs." From its fairly regular recitation of amateurs' contributions, the column offered a variety of new design ideas and, for example, how others handled the challenges of constructing and balancing a fully equipped telescope.

Also, since *Sky and Telescope* was always loaded with advertisements, I was envious of the fine accessories offered on its pages month after month, especially the occasional ads featuring clock drives from Ed Byers. With its gold-anodized look, a precision worm gear and a synchronous motor it certainly would save me the time and effort I spent attempting to craft my own geared drive. Accessories like those were beyond my reach at the time, but they certainly were a worthy goal toward which one might work. From then on I toyed with ideas of my own to fashion and outfit a reasonable mount.

I didn't work at this in a vacuum, as no one should. I sought out the books of the day that held the best information available. These references included "Making Your Own Telescope" by Allyn Thompson, and, the most respected of the lot, the three volumes of "Amateur Telescope Making" by Russell Porter and Albert Ingalls. In later years, I added Jean Texereau's tome, "How to Make a Telescope," and the more recent book "Telescope Optics" by Harrie Rutten and Martin van Venrooij.

Of course, this was more library reference than I'd ever need to complete a telescope, but, as I'd read and heard, the books held loads of fascinating detail, which is why I knew I wanted them. Of all of these, the three volumes of Ingalls' ATM books always have been proud additions to my bookcase. Digging through each of them every now and then as needed for some nuance of telescope-making always turned up a new, useful detail.

Early on, Thompson's book, too, was wonderful for understanding in words what Mr. Wolf was instructing at the Optical Shop. Though Thompson's approach truly was a homebuilt telescope from grinding and polishing the mirror to pouring hot tin alloy (aka "babbit") into pipe fittings to make the mount, the finished product was more than I felt I could attempt at the time.

Instead, humorous to me were several photographs in the book, first published in 1947. One of the photo plates in the front showed the author's finished product, a finely handcrafted six-inch Newtonian reflector on a classic homebuilt equatorial mount. To me, the comical part was the photo plate's caption, which claimed the entire scope, weighing in at 52 pounds, had been built at a cost of "less than

30 dollars"! The second photo plate on page 26, also amusing to me, was a scene of an optical shop, no doubt normal for its time, showing all of the men at work around their barrels, nearly all of them wearing neckties! One doesn't see such a formal approach to an avocation today.

I spent my own time around the barrel, most of the time not wearing a necktie, and mostly in the Optical Shop, as I built several more telescopes after my first one. I sought to have a representative collection of reflectors, and, in total, made several Newtonians, a Cassegrainian, and a Gregorian, all of different apertures, and nearly all with relatively short-focus mirrors. I wasn't attempting a certain quantity; I merely wanted a set of quality instruments. From my first mirror in 1971 to my largest mirror in 1996, I sought to have a usable set of optics for all my observing. By the mid-1990's, I had built those scopes in my spare time, like almost any ATM might've done.

Ultimately, I then did something different from most other ATMs. I used that last and largest mirror and set upon my project to build an uncommon telescope, not often seen in the public eye. In fact, I learned that the design was so singular that its identity had lain dormant for hundreds of years! To my knowledge, few have attempted the design – evidently I am among those few – but perhaps you, too, can add to their number.

For me, the fun for this project began while working on that last large mirror, a 17.5-inch glass from Coulter Optical. For what was forming in my mind, the mirror itself seemed large enough. Originally, it was a second-hand commercial mirror, already aluminized, which I'd purchased in 1986 from a married couple in Vandalia, Illinois, who were selling some of their son's hobby items for him to buy a motorcycle, so that he could deal with the narrow dirt roads of Zaïre, where he was to be sent by the Peace Corps.

But, after its purchase, I put away the big mirror until I'd start the large telescope in earnest. By 1993, I'd had the old aluminum coating stripped from the mirror by the glass firm of Clausing's in Skokie, Illinois, and then took it to James Seevers, director of the Optical Shop at the time, to examine its cleaned surface. (See Photo 3, pencil illustration of Jim Seevers.)

Photo 3

This drawing was made from a photo, shot in 1993, when Jim Seevers was Director of the Optical Shop, and, at the time when the author was working on the large mirror that would become the primary of a distinctive catadioptric telescope.

It was a moment of surprise when we placed the big mirror on the optical bench. Jim and I both had assumed that Coulter mirrors had a decent reputation for quality of manufacture, but were both startled and amused to see at the knife edge of the Foucault tester an odd expression staring back at us from this big mirror's face. Its surface showed hundreds of concentric rings, resembling an old vinyl phonograph record! The look of the whole surface indicated that the mirror had been apparently polished on a mechanical device, leaving the curious target of dozens of rings.

Superimposed on that surface was its optical figure, that is, an expression of its characteristic curve, which resembled the outline of an eight-sided stop sign. What we saw in terms of mirror optics was only a marginal, squared-off paraboloid. That is, the mirror's surface appeared as if it'd been polished minimally at a typical eight positions around the barrel and then sent on from their shop as complete. It didn't look as bad as my first dog-biscuited six-inch mirror from years earlier, and for that reason, Jim Seevers' prescription for that big glass wasn't the same as Ken Wolf's instruction for my first glass. Rather than have my mirror go back to grinding, this time the mirror appeared salvageable with careful re-polishing.

* * *

Does the ideal telescope exist?

Over the years, well after my first mirror was completed in 1971, I examined a number of designs for short-focus telescopes. Such designs included a variety of mounts, altazimuth and equatorial, and a variety of tubes, either open with struts or closed as a cylinder.

Every design I'd come to know showed some neat aspect of balancing the tube or allowing ease of access to the eyepiece. To know more, I'd have to see things for myself. One of the better places to see a variety of designs, up close and personal, is at a *star party*. If you're unaware of it, a star party is not a Golden Globes gala or a Hollywood night at the Oscars. Of course I'm not talking about movie stars. Think of a star party as a camping weekend with anywhere from 50 to 500 of your closest friends. It's a place where telescopes are aimed skyward all weekend, weather permitting. It's a place where families and singles, men and women, the young and the elderly, the hip and the geeky, all get together for clear nights under the stars. And there are plenty of telescopes, all makes and models, to look through and admire.

One of the better venues to attend is the star-party granddaddy of them all, *Stellafane*. Held outdoors every summer in the town of Springfield, Vermont, Stellafane does not boast to be a site perfect for dark-sky nighttime observation, but it does hold the distinction of being the best documented location for the origin of amateur telescope-making in the United States.

Stellafane was launched by Russell Porter – expert machinist, pencil artist, arctic explorer – after he founded the amateur astronomy group known as the Springfield Telescope Makers, who shared with colleagues his abiding interest in crafting unique and fine telescopes. (See Photo 4, pencil illustration of Russell Porter.) He was such an enthusiast for astronomy and telescope-making that he crafted as a Latin expression the name Stellafane, meaning "shrine to the stars." Located just outside of Springfield on a knobby summit known as Breezy Hill, Stellafane is readily identified by the steep roof of its pink clubhouse – yes, pink! – that he and his fellow ATMs built for their meetings and telescope-making. (See Photos 5, 6, and 7 for pictures of Stellafane.)

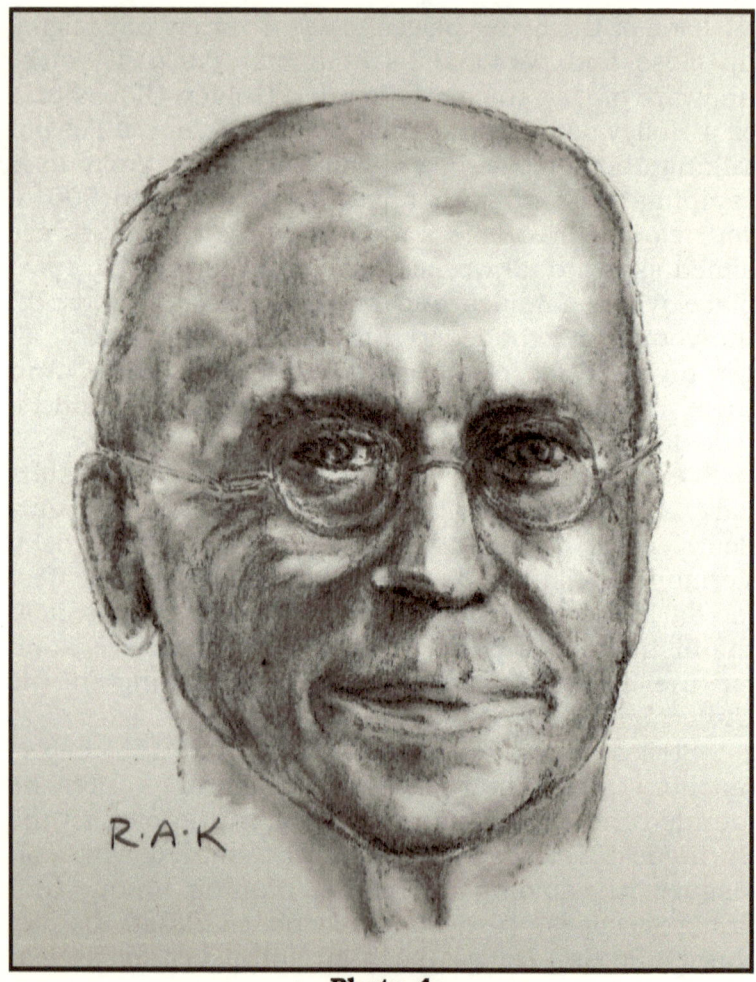

Photo 4

Russell Porter, founder of the Springfield Telescope Makers.

Photo 5

Long shot of Stellafane clubhouse, circa 1994.

Photo 6

**The author outside the Stellafane clubhouse
during a visit, mid-1990s.
Photo by John Mead.**

Photo 7

Interior of the Stellafane clubhouse.

The telescope designs of Russell Porter – that Yankee entrepreneur who knew how to make mirrors and craft telescopes – always intrigued me, not only for the obvious reasons that, in his writings, he encouraged and often made good use of scavenged materials, but that his common-sense designs often resulted in solidly mounted, workable instruments. From the original series of three ATM volumes edited by Albert Ingalls, what serious amateur doesn't know of Porter's attempt to craft a telescope mounting from an old engine block or, better yet, from concrete? (If you can't recall, check ATM Book One, pages 135 and 142.)

During the work on my later mirrors, as I gained practice with the Foucault tester and reading an optical surface, I also sought to follow some of Porter's example at using available materials with which to craft telescope mounts. I tried to salvage metal parts and wooden pieces, depending on the size of telescope tube, to make a decent mount.

Over the years I tried old camera tripods, but they were too flimsy. I tried a stout antenna mount – the kind used to install a TV aerial on a roof – but it shimmied too much. When I was still a teen, I even built a low-slung platform from thick wooden planks – left over from my dad's construction of a backyard picnic table – and clad them in cut lengths of surplus sheet metal – once used to line the concrete driveway – to build a heavy, stable platform, mounted on wheels from an old wheelbarrow. Truly it was a homebuilt affair. I kept it stowed under a work bench in my parents' garage to have it ready to roll onto the driveway for a night of observing. My dad called that contraption "The Tank."

Later, as an adult, I re-visited the tried-and-true technique presented by Sam Brown and made good use of iron pipe fittings, and even had bronze bushings inserted in place of interior threads, to make a smoother fit for steel shafts as axles for my homebuilt equatorial mount. The modified fittings were plenty sturdy, but the assembly was only useful for relatively short-focus telescope tubes, such as six- to eight-inch apertures. My intent was to have a suitable mount for a distinctive scope of relatively short focus, still fairly portable, but of far greater aperture than the mere six-inch glass I'd crafted in the Optical Shop.

This intended design presented some problems, especially if one wanted to use pipe fittings. The fittings would have to be overlarge, and I did consider their use. More to the point, a large-aperture scope meant that the entire instrument would not only be fairly hefty, but might be fairly tall. At the very least, it meant that one would have to climb a stepstool or a stepladder to reach the telescope's eyepiece, wherever the scope was aimed at the night sky. I always found that unappealing to do on soft, dewy ground at night.

Porter himself evidently had obviated the need to climb stepstools and stepladders with his own design of the *Springfield* mount. Named for the Vermont town in which he lived, the Springfield looked like an oddity but was a neat insight to a functional design that allowed the observer to remain seated while the telescope moved about his head. (See ATM Book One, pages 32-33.) No matter where the scope was aimed, one might have to straddle the central pier, but still could remain seated at ease. (See Photo 8, the Springfield mount.)

Photo 8

Porter's Springfield mount on display at Stellafane, Springfield, Vermont.

I became enamored of this concept and imagined the large-aperture scope taking shape in my mind as a likely candidate for the Springfield mount. I even began pricing the huge pipe fittings I'd need to order for telescope bearings and the thick-walled pipe for the central pier. What I had in mind, though it was still taking shape, was a design that allowed ease of access, a stable platform for astrophotography, comfortable seating for the observer, and no use of a stepladder. This is what I thought the Springfield mount might provide. But as marvelous as it seemed at the time, even Porter's ingenious design would not be the exact solution for the large-aperture scope I began to imagine. What I had in mind might sound like a dream telescope.

What I found, of course, is that no such thing exists. It is why I use this little book to report on the progress of a largely untried telescope design, one whose fundamental design is thoroughly distinctive and all done with mirrors. Please note this is not a book about making mirrors or building telescopes or even a draft set of plans for constructing an unusual optical configuration; rather, this book is intended to spark your interest in considering a workable design that helps to tackle a serious optical challenge.

One may wonder why a feasible design has remained experimental, even though its optical configuration does indeed allow easy access, comfortable seating, and no ladder. Not only for its novelty is it interesting, but worth acquainting more observers with this design for its obvious advantages and, in all fairness, for its potential pitfalls.

Of course, what I'll describe herein is no dream telescope, but I believe the design has considerable merit, either as a distinctive pursuit for the serious amateur who wants to enhance an existing high-quality refractor, or, as a scope worthy of an astronomy club or star party that wants to extend the accessibility of a telescope's eyepiece to small children and the wheelchair-bound. Sound unusual? Interested in knowing more? Please read on.

Chapter 3

Grinding Passion

A worthwhile goal, your first telescope

Some of the nicer people one might meet are amateur telescope-makers. Of course, that might sound like an obvious, self-serving remark, but indeed, I have met some very fine individuals, who, besides having a patient reserve which mirror-making requires, have a keen memory and a knack for detail; so, I've found my own observations to be quite true. For example, while working on my own humble six-inch glass at the Optical Shop, I first met Jim Seevers, working on a larger 16-inch glass. Years later, well after he became the director of the Optical Shop, I reminded him of our first meeting, and he was able to confirm in his own mind, with complete recall, by the mirror he once worked on, that we indeed had met!

Folks who stick with making mirrors and building telescopes have a certain demeanor, a calmness of attitude, and a world of patience. One has to have that, or eventually acquire these traits, because the art of mirror-making can tax one's tolerance.

I suppose one has to have a particular deportment to make a mirror and build a telescope, as it requires some degree of concentration and dedication. Grinding a mirror, for example, requires long hours, often working alone, if one hasn't the good fortune of a nearby optical shop to visit. Hours perhaps spent working at one corner of a stout bench along one wall of a basement, or on a wobbly bookcase if one hasn't a barrel.

For those who have never attempted to make their own telescope mirror, the procedure is fairly simple, almost primitive. Note the list of materials:

- Grinding barrel (or 55-gallon drum) as a table on which to work;

- Mirror blanks of Pyrex glass (6- to 8-inch diameter);

- Grinding abrasives (industrial grits from #80 to #320, and micron grits);

- Water in a plastic squirt bottle (tap water OK);

- Bucket, water-filled (or a sink with running water);

- Old newspaper;

- Small flashlight.

The actual procedure to grind a mirror sounds alchemical, but mainly, one uses just liquids and powders. Of course, the main liquid is water, since one constantly wets the surface of the glass to be ground. Every time one squirts water onto the mirror's surface, that's called a *wet*. A fairly obvious term, but, in the writings of traditional mirror-making, that's what it's called. On occasion, some ATMs have been known to use a few drops of glycerin added to the water, mostly in an attempt to keep the surface of the mirror from drying out too soon.

In actual mirror-making, the ATM attempts to grind two glass disks together in order to shape their surfaces to a desired curvature, which influences the eventual size of one's telescope. Although mirror-making books will outline their own preferences, the general recipe is straightforward.

Specifically, the mirror-maker starts out with two disks (called *blanks*) of Pyrex – the same glass of cookware fame – since that type of glass is hardened in its manufacture by double annealing, and, it resists changes in temperature better than plate (or window) glass.

One glass blank is placed on top of several sheets of opened-out old newspaper and in the middle of the top of the grinding barrel. This blank is also known as the *tool*, since it's the surface on which the actual grinding will take place. To keep the tool in place while grinding, the surface of the newspaper is wetted with water. This creates suction between the tool and grinding barrel, and keeps the tool from moving out of place. By the way, some ATMs like to affix rubber or wooden cleats to the top of the barrel, which very nearly immobilizes the tool for grinding.

The second blank is placed on top of the first. This second blank eventually becomes the *mirror*, and its surface, too, is ground. On the tool blank is squirted a small amount of water; a small amount of powdered industrial abrasive, called 80-grit, is sprinkled onto the water. The blanks are placed, one atop the other, that is, the mirror blank on top of the tool blank.

The suction of wet newspaper holds the tool blank in place. The mirror-maker places his hands on the mirror blank, and, with downward pressure, pushes the mirror across the tool. The mirror-maker will hear the harsh sound of abrasive grinding against glass (and that's good).

The mirror-maker keeps up this back-and-forth action for ten times. The mirror-maker then stops, rotates the mirror one-eighth turn to the right and then steps to the right one-eighth the distance around the barrel, and then resumes the back-and-forth action for another ten times. That is to say, if one steps counterclockwise around the barrel, the mirror rotates clockwise under your hands. Or vice versa. By the way, I've noticed that whether one writes left-handed or right-handed may influence which direction one steps around the barrel. Nevertheless, the direction of the mirror's rotation should be opposite to the direction one steps.

Mirror rotation and stepping around the barrel go hand-in-hand with the downward pressure of grinding. Approximately 20 pounds of downward pressure is applied. For an example, pressing your hands down on a bathroom scale will give you the actual feel of 20 pounds of pressure.

A total of eight changes of position, or one complete turn around the barrel, will suffice to provide even wear over the surface of glass being ground by the mirror-maker. When the grinding begins to sound less harsh, it's time for another wet. That is, the mirror-maker stops to re-charge, that is, to add more of the grit and water between the contact surfaces of both blanks.

The push against the glass tool should not exceed half the diameter of the mirror; otherwise, the mirror will teeter and drop off the edge of the tool, which could chip either glass blank. Even though they're made of Pyrex, the blanks can still chip along the edge. One should file, or *chamfer*, completely around the rim of both the mirror and tool to a 45-degree angle with an Arkansas stone or similar gritty grinding plate. The action of chamfering will lessen the chance of chipping at the mirror's edge.

By the action of grinding by hand, the mirror-maker eventually hollows out a shallow (concave) curve in the mirror blank on top of the tool blank, while the tool blank becomes slightly dome-shaped (or convex) in response to the hollowing of the mirror blank. The mirror-maker can test the degree to which the curve is made concave on the mirror blank. As the curve deepens, successively finer abrasives are used to smooth the curve.

The industrial grits are labeled generally from #80 to #120 to #220 to #320. The higher the number, the finer the abrasive; generally the number means the numbers of individual grains of abrasive per inch. So #80 is for rough grinding, at 80 grains per inch, while higher numbers are for finer grinding and smoothing the roughed-out surface.

When the desired curve is achieved with rough grinding, both tool and mirror are dunked in the water-filled bucket, or rinsed with running water, and cleaned of all grit. Finer abrasives (namely, #120, then #220, then #320) are used successively to smooth the curve generated by grinding initially with 80-grit. Ever finer abrasives, of several-micron grain size, are used to smooth and finish out the overall fine grinding of the mirror's surface.

The grits are commercially made, used in industry, and are non-toxic. Rougher grades of abrasive, mainly silicon carbide, have grains with microscopic sharp edges; finer grades of abrasives, mainly aluminum oxide, have rounded edges. The rougher grades of abrasives will scratch both glass and plastic surfaces, and can roughen skin and dull fingernails. The finer grades of abrasives simply turn to a thin mud during a typical wet. By the way, a stationary tub in the basement or slop sink in a utility room might be best to rinse off grits, but can, just like sand washed down a drain, accumulate in household plumbing.

Wetting and charging, pressing and grinding, rinsing and drying are the steps to working a mirror through the stages of shaping and smoothing its curve. Grinding a mirror through its set of abrasives can be messy when you start, but gets less so as you move to successively finer grits. When you read about grinding a mirror in the ATM books, the descriptions are various but uncomplicated. Wetting with water the surface of a disk of glass and charging with a dash of grit is all that's needed to start the scratchy, raspy snarl of mirror and tool together. For every step you take around the barrel – eight steps to make one complete turn – another one-eighth spin of the mirror in the opposite direction ensures, with good pressure, an even grind of the entire mirror's surface. That's essentially the grinding procedure.

Staying with this same, simple recipe eventually can hollow out a shallow curve in the upper disk of glass, the mirror, and a slightly humped surface for the lower disk of glass, the tool. The cutting action of various grades of industrial strength abrasives, or carborundum, is quick enough to see demonstrable results within a matter of a few short hours, but slow enough, even if the mirror is made of doubly annealed Pyrex, to keep the novice glass pusher out of trouble.

The emerging optical surface on the mirror can be tested with a small flashlight aimed at its cleaned, wetted surface to determine the mirror's *radius of curvature*, that is, the size and shape of the slight curve represented by the mirror's surface.

To understand the actual shape of the surface of a telescope mirror, one has to visualize, in three dimensions, the meaning of radius of curvature. Think of the round surface of the ground face of that emerging mirror as one small segment of an enormous sphere, a big imaginary ball, suspended in the same room with you. The radius of that big, phantom ball represents the mirror's radius of curvature. Another way to see it is if you're standing at the radius of curvature, you are at the ball's exact center and your mirror's concave surface is part of the ball's big round imaginary surface in space.

Hold that small flashlight next to your eye and then shine its light at the mirror's surface, wetted with water, and you'll see a reflection of the light source. A slight motion of the flashlight back and forth next to your eye will show on the wet surface of the mirror whether you're inside or outside the mirror's radius of curvature.

If the reflection moves with the motion of the flashlight, you're inside the radius of curvature; if the reflection moves opposite to the motion of the flashlight, you're beyond the radius of curvature. Also, the more finely ground the mirror's surface, the clearer the reflection of the light, and the less need to wet the mirror's surface with water, which is why a flashlight works well enough early on and a small bulb is just as useful later.

If the light's reflection blooms or brightens all at once, you're very near, or at, the exact radius of your mirror's surface. Measure off the number of inches (or centimeters) that both your eye and light are from the mirror's center, and that's your radius of curvature. In an optical sense, half that value is your mirror's *focal length*, the distance from the mirror's surface to a point in space where the mirror concentrates light to an image in crisp, clear *focus*. The number of mirror diameters it takes to equal that focal length is known as your *focal ratio*. This measurement and its attendant arithmetic (learn it well!) is basic to charting your mirror's progress through both rough and fine grinding.

The usual, safe focal ratio for a novice glass-pusher is $f/8$, which means that the telescope mirror's focal length is eight times the mirror's diameter. This makes that usual four-foot tube so easy to carry. But I wanted a shorter tube, so the mirror needed a trim ratio. Mine was a measured focal ratio of $f/5.7$ – originally I'd aimed for a nice-sounding ratio of $f/5.5$ – purposely short to make it compact and easy to transport into the field.

To chart one's own progress through the art of mirror-making – it's a good idea to keep a log book – takes more than a rough estimate of the time and effort you'll need to accomplish the task. It takes a fine determination to complete the project, not only to one's own satisfaction, but to others' inspection. Make a good telescope and you achieve a personal sense of accomplishment. Build a fine telescope and you inspire others to build one just as good or better. That's the determination I made for myself when working toward my goal to complete my first mirror.

Prior to working on my mirror at the Optical Shop, I simply wanted it done. I followed the foregoing recipe and thought I'd get a fine result. In hindsight, that might've appeared hasty, and for that hastiness, I got a bad mirror. Having someone point me in the right direction, and literally send me back to grinding, taught me to correct my errors and not to rush the work.

Taking time to assess, taking stock of one's status, then easing into the rhythm of the art is what makes one avid about his task at hand. The task then becomes, not a chore or a burden, but almost a duty, and one that takes on an ease to achieve. It's the quiet, deliberate determination to complete it well that rewards one with a mirror worthy to use and perhaps be the envy of others.

* * *

A worthwhile design, the Mersenne-Nasmyth

The larger telescope I began to imagine eventually became clearer after a fortuitous visit to Stellafane in 1994. At that annual summertime meeting of telescope-making enthusiasts, I came across an odd design that seemed to marry a fine refractor with a large reflector. I didn't think this was possible and lingered near the scope to learn more about it. The larger scope itself looked usual enough in its classic Newtonian configuration as a Dobsonian; however, angled up from one side of its altitude axis was a high-quality refractor. Seated behind that refractor was the man who built the whole scope, but he seemed reticent to talk about it. He called his scope's design a *Mersenne-Nasmyth*.

Named for two important historical figures, Marin Mersenne and James Nasmyth, this design wasn't well known in the popular optics literature. Various individual abstracts in university publications referred to it obliquely; more often, it likely might be classified as the design one would want to avoid. There's some small reason for that, but more on that later.

Marin Mersenne, a Franciscan friar, was a French mathematician of the 17th-century and an ardent champion of Galileo and his discoveries. He attempted to obviate the need for lenses in a telescope with his own two-mirror designs, but his good friend, René Descartes, talked him out of any further pursuit of optical configurations. Instead, Mersenne is better known for experiments in tuning musical instruments and in music theory, and, for his *Mersenne primes*, a special category of prime number important in solving complex computational problems.

Mersenne detailed his distinctive optical designs in his own book (*L'Harmonie Universelle*, published in 1636), describing in drawings the actual ray traces of two, distinctive, though untried, two-mirror configurations. However, his optical designs never caught on within his lifetime. More to the point, Mersenne himself was unaware of the subtle, advanced significance of the optical designs he'd concocted. This is evidenced from his erudite correspondence – Mersenne was an indefatigable letter writer – with Descartes. It would be a long time until others would explore the sophisticated utility of his work.

James Nasmyth, a resourceful toolmaker, was a 19th-century Scottish engineer and inventor, known for his development of the steam hammer, which quickened the manufacture of metal forgings. He dabbled in astronomy, and chiefly liked viewing the Moon. He even co-authored – with James Carpenter, a British astronomer and his contemporary – a detailed book about it. Nasmyth also developed a type of telescope that afforded him a comfortable view of the night sky while seated next to his instrument. For example, by inserting a flat mirror in the light path of a typical Cassegrainian design, he diverted the light path through the declination (or, altitude) axis of his telescope. This allowed a steady and convenient place for the eye to observe. This innovation, known as the *Nasmyth focus*, is used currently on large telescopes in large observatories.

Clyde Bone was the first person I'd known to construct the Mersenne-Nasmyth telescope, and he was the owner of the distinctive telescope I saw at Stellafane. I wasn't familiar with the design at the time I met Mr. Bone, but his scope certainly made an impression on me when I returned that evening to look through it. Aimed at the Andromeda Galaxy, the scope's view of our nearest big spiral neighbor was nothing less than stunning. The galaxy filled the field of view impressively and with pleasing detail.

Mr. Bone's telescope was a short-focus construction, but still relatively tall with its 20-inch diameter mirror of $f/5$ focal ratio. Odd to its construction was that it was a telescope of dual perspective, that is, a combination of *two* telescopes. It was both a Newtonian reflector *and* a Mersenne-Nasmyth design, and fairly easily switched from one mode to the other. In Newtonian mode, one might need a stepstool to reach the eyepiece, but the Mersenne-Nasmyth mode only required a chair on which to sit. And Mr. Bone demonstrated that well. (See Photo 9 of Clyde Bone and his signature telescope.)

Photo 9

Clyde Bone and his unique Dobsonian-style telescope, a scope of dual perspective, convertible from Newtonian to Mersenne-Nasmyth optics.

After Stellafane that year, I tried to look up aspects of the design and did not find much about it, though Mr. Bone himself had already written an article about his 20-inch aperture design that appeared in *Amateur Astronomy Magazine* (#4). As I understood from his article and scant details in the optics literature about this design, his original configuration used a system of *confocal* mirrors, that is, complementary curves between concave and convex optical surfaces. A subsequent article about this design, also written by Bone, appeared in *Sky and Telescope* (September 1999, p. 130), where, after constructing his first 20-inch aperture telescope, he described a larger version of his Mersenne-Nasmyth with a 30-inch aperture. By his own admission, incidentally, his 20-inch and 30-inch apertures were the only two telescopes he'd ever built!

According to Mersenne himself, it was possible to construct at least two different optical designs using a configuration of confocal mirrors. Though he never actually constructed either configuration, he drew out the precise arrangement of mirrors to complete his designs. He proposed that the result of either of his configurations would result in a collimated optical path, namely, a magnified, bundled beam of light. Mersenne was correct. In fact, Mersenne had invented an optical system with telephoto traits.

Though he didn't recognize its entire significance at the time, in theory and in practice, this distinctive design becomes a beam compressor for the light path. That is, the virtually parallel rays (from, say, a star) entering this optical system are compressed, magnified, and remain parallel on exiting the optical system. The advantage is that the compressed light beam can then be further magnified.

The use of additional, high-quality, flat mirrors allows that light path to be directed into a small telescope, usually a first-rate refractor, at a Nasmyth focus, i.e., through the declination axis, or, the altitude axis of the telescope. Happily, one can then enhance the use of a quality telescope of small aperture, making it behave like a larger instrument!

As a hybrid arrangement, this optical configuration combines the magnification ability of a high-quality refractor, technically known as a *dioptric* device, and the light-gathering power of a large reflector, known as a *catoptric* device, resulting is a true *catadioptric* telescope, the name of which suggests a true marriage of two different optical designs.

Subsequent to Mr. Bone's initial work, I myself attempted this design. I'd worked on mirrors for other telescopes during this time in the Optical Shop, but by 1998, I'd returned to the Mersenne design in earnest, constructed a prototype of plywood – nick-named *Styckehenge* – and found it both initially workable and a worthwhile endeavor. More important, it led me to a better understanding of large-telescope construction and to the original design's good and bad points. (See Photo 10, of the author polishing the Coulter mirror.)

Photo 10

Here the author polishes the 17.5-inch Coulter mirror at the Optical Shop, circa 1994. Photo by Jim Seevers.

The optics for *Styckehenge* not only included the re-worked Coulter mirror, but needed additional mirrors to complete the configuration. Once the main mirror was re-polished and re-figured, I set to work on its confocal match.

To make that match, it was evident from the scant literature available to me then that a Mersenne-Nasmyth was an offshoot of the classic Cassegrainian design, which means that the secondary mirror, as in Mr. Bone's design, would have to be *convex* and match the same optical figure as the *concave* primary mirror. With that mind, I simply used the same arithmetic presented in ATM Book One (for calculating a Cassegrain) to figure the likely size for the secondary mirror in my emerging Mersenne-Nasmyth optics.

In this instance, a confocal match means that the secondary mirror has the same focal ratio as the primary mirror. So, for my re-worked Coulter mirror at a focal ratio of $f/4.5$, the secondary's focal ratio would have to be exactly the same. Once again under the guidance of Jim Seevers at the Optical Shop – grinding, polishing, and figuring a 4-inch diameter mirror on its glass tool – I completed the convex secondary in a timely manner. Now I had the two mirrors, concave primary and convex secondary, with which to use in crafting the Cassegrain version of this unique, confocal design.

After its completion, though, what became immediately evident was that I also had another secondary mirror. Yes, I was readying a *convex* secondary as a confocal match for a Mersenne-Nasmyth that was essentially a variation of a Cassegrain design, but, in the leftover glass I now had another confocal match, a prospective *concave* secondary. This was intriguing because that leftover glass potentially formed the basis for Mersenne's other confocal configuration, namely, that of a Gregory design! So now I again had two mirrors, concave primary and concave secondary, with which to craft this other one-of-a-kind confocal design, just as Mersenne himself had described more than 350 years ago.

I learned later that several online threads discussed the prospect of the Gregorian alternative (as found in the Online References at the end of this little book), but, apparently at the time of those online threads, few knew wholly of Mersenne's valuable contributions to modern optics. In the meantime, it was for me an important departure, and, in a direction for a promising chance at an intriguing optical configuration.

Where I could make a workable Mersenne-Nasmyth design, I found also the potential for a feasible Mersenne-Gregorian design. Of course, the optical configurations would be different from one another – still I saw them as *complementary* designs – just as Mersenne had shown. The switch from a Mersenne-Nasmyth configuration to a Mersenne-Gregorian configuration seemed a natural segue, that is, a paired design of one to the other, effectively of a Cassegrain to a Gregory. And, evidently, as configurations untried by telescope-makers until the 20th-century, that was exactly the conclusion Mersenne had reached in the 17th-century.

It is noteworthy at this point to reference the more recent works that help to highlight the Mersenne variants. Two of these references were not available at the time when I first began to explore the Mersenne design after I met Mr. Bone. In the reference list at the end of this book are two excellent books – one by Gérard Lemaitre, the other by Ray Wilson – on all types of optical designs, including Mersenne's configurations. What these authors conclude is that Mersenne not only had introduced workable optical configurations, but didn't realize just how advanced they'd prove to be. Originally, what

Mersenne had described in his book were *afocal* versions of the optical designs of Laurent Cassegrain and James Gregory. That's essentially the nature of Mersenne's designs. One is a complement, a variant, of the other, as Mersenne had first described them.

What started as a test, then, to see whether I could even replicate Mr. Bone's success in crafting a distinctive telescope, inspired me to plan also for that complementary design. My plywood prototype, in the guise of *Styckehenge*, became that test model for Mersenne's own variations.

First, the completion of that first prototype introduced me to the construction of the classic Dobsonian, an approach to telescope-making inspired decades ago by ATM John Dobson, who pioneered the assembly of large Newtonian optics built boxy and low to the ground for stability while introducing relative ease to their transport. (See Photo 11, *Styckehenge* at a star party.)

Photo 11

A photo of *Styckehenge* at a star party where it debuted and received a design award.

Next, since my own Dobsonian version of the Mersenne-Nasmyth resembled a large, square barrel, it allowed room inside to store the rest of the scope as nested components and to gain simple access to the main mirror. Essentially a plywood rocker box supporting a cage of rings atop aluminum struts, the important point is that the optical design was versatile enough, allowing me to replace easily its convex secondary with a concave secondary.

Then, by adding in struts of nested conduit, I could extend the distance between the primary and secondary just enough to have the nominal workings of a Mersenne-Gregorian design. (See Photo 12 of extended struts to make a Mersenne-Gregorian.) So, with a few simple adjustments between designs, I had the makings of Mersenne's own original afocal configurations, that of a Cassegrain and that of a Gregory.

Photo 12

The *Styckehenge* scope shown in extended Mersenne-Gregorian mode.

The Mersenne-Nasmyth, i.e., the afocal Cassegrain, and the Mersenne-Gregorian, i.e., the afocal Gregory, both have a secondary which is confocal to the concave primary, just as Mersenne had proposed. The compressed beam of light produced by either of these optical configurations produces a magnified beam of parallel light rays, namely, a telephoto effect, which can be fed into an accompanying well-built, smaller telescope. More to the point, not just any small telescope, like a compact Maksutov or Schmidt-Cassegrain, but a decent, high-quality refractor. That is to say, in modern folded designs, the mirrored telescopes are mismatched to a confocal configuration since they introduce too many optical defects to the light path, whereas a well-made refractor receives a field of curvature with minimized aberrations. More on this later.

This first set-up of *Styckehenge* was admittedly crude; nevertheless, it did demonstrate the basics of both the Mersenne-Nasmyth and the Mersenne-Gregorian designs. I learned much on how to refine the construction of these designs, and subsequently set about trying a different assembly of the optics, still movable, eventually heftier, but hopefully even more stable.

And, to my surprise and delight, I discovered for myself the brilliance of Mersenne's own work as a rewarding alternative to conventional optics.

Chapter 4

Smoothing the Curve

Mirror-making becomes rewarding

My own first mirror was fairly steep for a first-timer, so its concave surface kept me close in when examining it, first with a flashlight and a wet surface during rough grinding, and then with a large Christmas tree bulb and a dry surface during fine grinding. With a focal ratio of $f/5.7$, my six-inch glass needed that extra care to make certain it was free of pits out to the very edge with a smooth curve.

The edge got its share of work, so much so I ran the risk of a "turned" edge, or more accurately a "turned-down" edge, a common defect. Fortunately, I was able to avoid that hazard with Ken Wolf's expert instruction. Since my aim was a short focal length, the mirror had to have a relatively short radius of curvature. To get even coverage of grinding and still maintain a smooth, relatively steep curve, there are various techniques the glass-pusher must use. If one simply moves the center of the mirror on top over the center of the tool on the bottom, the curve remains fairly hollow and the action is slow. Center-over-center strokes are fairly safe for the novice. However, if one hangs the mirror to one side so that its center rubs over the edge of the tool, the action is faster near the center of the mirror, deepening its curve.

This latter stroke I used most to achieve a short focal length, and for me it worked. Too much overhang runs the risk of chipping the mirror if the glass rocks the mirror off the edge of the tool. Chipping the edge of a mirror might not matter on its back, but, in my experience, chips along the edge come more from the slightest side motion than from any accidental rocking or knocking the mirror. A glass edge's typical conchoidal fracture, resembling the shell of a mussel, occasionally made me feel I might have better luck knapping flints into arrowheads than grinding telescope mirrors.

It was a clarifying moment in fine grinding when the last vestige of tiny, visible pits rimming the very edge of my six-inch glass disappeared for good. I recall realizing that I had fully removed that old dog-biscuited surface for good, and that now I could move into the polishing room at the Optical Shop. It was a fine feeling of satisfaction for a young teen moving closer to his dream of crafting his own telescope.

* * *

The Mersenne-Nasmyth offers a rewarding alternative

In an attempt to improve the overall usefulness of the Mersenne-Nasmyth telescope, I opted for his alternate complement of mirrors, namely, two confocal, concave optical surfaces. As expected, a collimated light beam was the same outward result. I verified this particular configuration with ray tracing software available from publisher Willman-Bell. (Check the list of Online Links at the back of this book.) Since that prototype resembled a longer Gregorian-style system, it was rightly an attempt at a Mersenne-Gregorian system, which remains a feasible, but largely untried, design. That alone made it worthwhile to try. And, again, it was exactly what Mersenne himself had described.

From my perspective, two clear benefits were erect images – typical of Gregorian optics – which makes the public's viewing of extended objects – like the Moon – easier to interpret, and, the system's relative portability. My article in *Amateur Astronomy Magazine* (#39) suggests advantages to this design.

In the Mersenne-Nasmyth system, the concave main mirror has a small convex secondary as its match. That is, both mirrors have the exact same focal ratio. By contrast, in the Mersenne-Gregorian system the main mirror has a small concave secondary, which, too, matches the primary in f-ratio. In both systems, the optical surfaces of the primary mirror and of the secondary mirrors, whether convex or concave, are all paraboloids.

On occasion, I've been asked whether confocal spherical surfaces might work just as well as parabolic surfaces. Confocal optics can be tricky in that one must achieve a finely figured match of optical surfaces. (There are several links listed in Online Links at the end of this little book that discuss alternate optical figures for confocal mirrors.)

The purpose of either the Mersenne-Nasmyth or the Mersenne-Gregorian optical arrangement is to deliver as much light as possible to a receiving telescope. Even though neither system in a confocal configuration brings light rays to a focus – parallel light rays entering the system remain parallel on exiting the system – one still needs the correct optical surface to concentrate the most light, and the parabolic surface of a mirror does that best.

Now let's see if an innovative Mersenne scope is any better than traditional designs.

Chapter 5

Keep Focused

Are we there yet?

On setting up on a barrel in the polishing room of the Optical Shop, I felt I'd arrived at amateur telescope making. My finely ground mirror looked very good, and I was prepared to make it even better. But I wouldn't be using the same items or procedure that led me to a dog-biscuited mirror my last time up at polishing. This is where expert instruction and a new approach came into play.

Since glass doesn't polish glass, the art of mirror-making has required a suitable surface on which to remove the microscopic pits leftover from fine grinding and to impart a clear-as-glass shine to the whole of the mirror's surface. For this, tree resin, viz., *pitch*, is heated, melted, and deposited to form a layer – a *pitch lap* – on the tool, becoming the surface against which the mirror is polished. To improve the soft bite of pitch and facilitate even polishing, the layer of hot resin is often formed with minute channels or grooved in some manner.

At the Optical Shop, gone was the need for the typical pitch lap outlined by Sam Brown and Russell

Porter, since its opticians had different methods. Gone was the need to line the tool's edge with a paper collar to dam hot pitch from spilling, since practitioners of polishing at the Optical Shop literally turned the art of making a pitch lap upside-down and poured hot pitch instead onto the mirror. Gone also was any use of a square-cut mat or, worse yet, any attempt to cut channels through cooling pitch.

At the Optical Shop, the opticians instead used a hex mat, i.e., a flat rubber honeycomb, to form hexagonal channels while the mirror was still warm. Gone, too, was the traditional use of a polishing agent known as jeweler's rouge, since two commercial products, one known as Barnesite, an orangey brown powdered mix of rare earths, and cerium oxide, a whitish powder watered to a slurry, were the preferred polishing agents.

For pouring a suitable pitch lap at the Optical Shop, a mirror, concave side up, was heated in warm water, dried, then quickly painted with a thick mix of Barnesite and water. The hex mat, also completely layered and juicy with the same mix, was laid off-center atop the coated mirror. Hot pitch from the pitch pot was then dispensed from a small spigot to cover just the area around the center of the upside-down mat and mirror.

Then the glass tool, also heated and dried, was pressed, face down, onto the hot pitch. Even pressure ensured that the hot pitch squeezed out to the very edge of both tool and mirror. The whole sandwich of warm glass, honeycomb rubber, and hot pitch was flipped right-side up and then pressed down further into place. The mirror on top was then evenly slid off the hex mat and set aside. Before the pitch cooled too much, the hex mat was peeled up gently from the pitch in one even motion. Too warm, and the pitch left long strings like spaghetti stuck to the mat; too cool, and the hex mat chipped its way out of the pitch when pulled. Either extreme condition might mean patching the lap with hot pitch. Too messy. Or, one could simply start over. A short wait while the pitch cooled, then one took a hammer (yes, a hammer!) to chip away the cold pitch, gather it up, and warm it again in the hot pitch pot.

Once the pitch lap was readied, the ATM resumed his craft, but with a different emphasis on stance and stroke. The previous eight positions around the barrel remained standard, but standing still stayed in the grinding room. As a practitioner of polishing, the ATM now had to put some whole-body back-and-forth muscle into his stroke, and the stroke changed as well. To get the effect of maximum effort, the practitioner practiced a long stroke but ran the risk of possible defects at the mirror's edge. With good guidance, a variety of short, long, and occasional overhang strokes delivered precise polishing right out to the beveled edge of the glass.

Polishing can be arduous and lasts for many hours. It's not necessarily physically taxing, just repetitive with lots of elbow grease. One must stand with feet apart on the floor to affect a stance that will control the action of polishing on top of the barrel. Downward pressure is still applied, but no fingers dangling over the mirror's edge. Occasionally, the ATM could speed the polishing process by flipping his sandwich of glass and pitch. A pitch lap on top can cut quickly, but can also wound one's pride more deeply if the action cuts too fast. Then you might have to backtrack and devote time to correcting a flawed surface. Don't hurry; take time to get it right.

One works toward crafting a precision surface and either hasty work or slovenly ways can defeat the process of polishing. That may sound unkind, but ATMs familiar with historical mirror-making know the

demanding lengths to which mirror-makers have gone to ensure a precision surface free of defects. (See ATM Book One, pages 80, 280, 291-294, and 299.)
That precision surface can take on a variety of appearances throughout polishing, which is not at all unusual. Knowing how to interpret that appearance is the key to eventual success. For example, do not think for a moment that the bright gloss one usually finds after only an hour or two of polishing is the finished product. Far from it. Initially, some hard-charging in polishing stance can impart a clear sheen to most of the surface, but that's merely a "false" polish, a superficial effect. Don't be fooled. Complete polishing takes a few to many more hours, depending on the size of the mirror, good contact with well-fitting pitch lap, constant and consistent action, tidy ways, and patience beyond measure.

* * *

How is a Mersenne scope any better?

Since a Mersenne-Nasmyth optical system is an offshoot of the classic Cassegrainian, one must be patient with its idiosyncrasies. For example, the curvature of field created by the optical system does not allow it to be used with all eyepieces. Put simply, individual eyepieces differ by their field curvatures, due to their individual designs.

Through trial-and-error, then, one can find the best match of commercial telescope and eyepiece to match the curvature of field created by the Mersenne-Nasmyth system, since this optical system (like the classic Cassegrainian) is *concave* toward the sky (that is, inward curving). By comparison, the Mersenne-Gregorian system also presents a similar challenge for field matching, since it too has a distinct curvature of field, but is *convex* to the sky (that is, outward curving) and so is opposite a classic Cassegrainian. By matching field curvatures of either Mersenne configuration with the field curvature of selected eyepieces, the resulting image in the eyepiece's field of view ideally should give the appearance of a relatively flat field.

From the scant literature on the subject of Mersenne optics, there emerges a neat detail, namely, that since the distance between the primary mirror and the secondary mirror is so great, a Gregorian variant may have inconsiderable optical defects. That is, the extended distance between confocal components may essentially trivialize any other optical defects inherent to a mirror system. This sounds promising to those who might consider this optical design.

However, according to this same description in the literature, the resulting magnification and narrow field of view from the Gregorian variant of a Mersenne might also be very near to the effective maximum for such a system, rendering it useful for only limited night-sky work. In other words, the development and deployment of a Mersenne-Gregorian design, operating within critical tolerances, might make it practically useless for any flexible night-sky observation.

I certainly don't agree with that assessment, since the only contributors to this viewpoint evidently have not built a working model of either a Mersenne-Nasmyth or a Mersenne-Gregorian telescope. I have come to know first-hand the promise, perils, and pitfalls of this design. Yes, either optical configuration has its idiosyncrasies, but I can say that the overall approach to either the Mersenne-Nasmyth or Mersenne-Gregorian design is one of promise, which far outweighs the potential perils. Also, I can say confidently that neither the Mersenne-Nasmyth nor Mersenne-Gregorian design has any more idiosyncrasies than any other folded optical configuration. At present, a 17.5-inch $f/4.5$ paraboloid is my main mirror. When used in Mersenne-Nasmyth mode, its confocal *convex* match is four inches wide and placed at a calculated distance of five feet from the primary. When used in Mersenne-Gregorian mode, its confocal *concave* match is placed eight feet from the primary in Mersenne-Gregorian mode. A tertiary flat just above the primary directs the collimated beam into a 94-mm $f/7$ apochromatic refractor from VernonScope USA.

That sounds straightforward enough, but one must consider the implications of either design. Exposed secondary and tertiary mirrors, that is, small mirrors having nothing to shield them, can lead to unwanted stray light, known as *off-axis* light rays, hence the likely need to install a *baffle* around each mirror. Also, unbaffled components gives the Mersenne-Nasmyth and Mersenne-Gregorian its most baffling idiosyncrasy, that of a superimposition of a "high-power" view over a "low-power" view. More on this later, but the quick fix is that both components ought to be baffled to eliminate this quirk.

The best overall telescope to place at the Nasmyth focus for either Mersenne-Nasmyth or Mersenne-Gregorian design is a high-quality refractor. A classic sky-blue Brandon refractor serves as my receiving scope and has a large, high-quality, quartz diagonal to place the image at a comfortable position for most seated observers.

In effect, the f-ratio of this enhanced system becomes the f-ratio of the receiving refractor. Though I use an $f/7$ refractor, ideally an $f/4.5$ model might be superior, since the f-ratios of all inherent components (viz., primary, secondary, and receiving refractor) ought to match. By the way, this is the principal assessment I learned from Clyde Bone. (Again, please see his article in *Sky and Telescope* magazine, September 1999.)

The platform on which the refractor rests is aluminum plate and allows the Brandon to slide in and out on its own dovetail for relatively easy set-up. The whole system swings smoothly in azimuth on a rebuilt stout tripod (manufactured by Houston Fearless). The entire assembly can be pulled fairly easily on a half-dozen pneumatic wheels.

Next let's see what it takes to make the Mersenne design work.

Chapter 6

Go Figure

That wasn't there a moment ago!

Across the surface of one's mirror during polishing, hills and valleys, peaks and canyons, mark the sub-microscopic elevations of this glassy landscape. It is the challenge of the insightful ATM to become familiar with the lay of the land and then tame this wild countryside.

In ATM parlance, *figuring* is the careful polishing that gets one to the best surface for the mirror, and the *figure* is the optical landscape seen at the knife edge of the Foucault tester at any given time. Ideally, every ATM aims for a paraboloid – a parabola in three dimensions – but likely passes through a spherical figure, then perhaps an ellipsoid, or even a hyperboloid, on their way toward the desired parabolic figure. Recognizing and reading these figures with some degree of ease is the province of the expert optician.

In the course of preparing that first glass at the Optical Shop, I was guided to face as few optical figures as possible. That was the experience Ken Wolf brought to his instruction in mirror-making. At the

Foucault tester, it was plain to see, time and again, once one knew how to interpret the mirror's shadowed geography, the glassy landscape being wrought by one's own two hands could be navigated fairly smoothly with the right guidance.

A turn or two around the barrel with the correct polishing stroke could erode an optical peak that might represent an epoch of geologic time, while an untoward stroke could scar the landscape for what might seem like a good deal longer.

Perhaps the most feared feature – and the one most commonly seen in mirrors among novices – is the "turned" edge, an optical defect at the outer rim of the mirror's circumference. This feature does not mean that the mirror's edge has morphed into something hideous – though the novice's reaction to a turned edge might make it seem so – but that the very edge of the mirror appears like an optically steep cliff. In reality, it's a flattening at the edge of the mirror's surface, lengthening the radius of curvature for the mirror's edge. It just looks like a precipitous drop-off at the knife edge of the Foucault tester. It is one of the more difficult features to correct, since its presence suggests planing down all of the surrounding foothills that rise up to place that cliff at the mirror's edge in the first place.

Every optical figure, every sharp or subtle shadow, that can be viewed and interpreted with the Foucault tester can be explained. And the resulting figure is always due to the person polishing the surface. If something goes awry, there is no one else to blame (unless someone unseen does something sneaky to your mirror). It should be obvious enough that an optical figure, good or bad, is directly due to the efforts of the ATM that polishes it.

Sir Isaac Newton, who invented and built his own working version of the reflecting telescope, understood not only optics and telescope-making but the forces that shape the surfaces his laws of motion describe. One must take time to grasp what happens at the surface of the mirror. If the ATM bears down with too much force in his haste to complete the polishing of the mirror's rim, a turned edge can result. If one exerts too little force or strokes wildly in the wrong direction, a central peak or a grand canyon may suddenly appear. If the figure changes, and does so often enough, it's not the laws of physics that are changing; rather, it is the novice who needs to change his errant ways.

Growing up with the encouragement of friends with interests in mirror-making was especially rewarding to keep one going and to keep one from errant ways. At the time I was a sophomore in high school and still an adolescent at puzzling out how to read shadows from a Foucault tester. Two acquaintances in high school – upper classmen at the time – worked on their own mirrors and were successful in making their own telescopes. Showpieces those scopes were and good examples of what steady, mature workmanship could accomplish. One was a complete homebuilt Cassegrain of pipe fittings; the other was a classic Newtonian along the lines of Allyn Thompson, but completed with machined (not babbitted) bearings made by my friend's father, who was employed at the time as a machinist at a Chicago university.

While these were finely crafted affairs, I myself still tinkered with pieces of scrap with which to fashion a tube and a mount. I meant to learn and do this on my own. So I attempted various designs and tried to make do with what was available, which didn't often make for a very sturdy tube or mount. Nevertheless, I continued to learn about telescope design and the materials with which to build them better.

In high school, when I used to visit the Optical Shop, usually on Saturdays, I became acquainted with fellow ATMs and learned tips and tidbits about the construction of telescope tubes, mirror cells, bearings, and mounts. Such fun conversations helped to bring to life the discussions on the pages of the few books available at the time. It seemed that such talk could hold my interest forever. Of course, one of the fine people I met back then was Jim Seevers, already an accomplished optician on his own and making mirrors for others, who one day would become the director of the Optical Shop.

It was such a fun time to learn what likely could work, and perhaps what might not, both in the art of mirror-making and from the science of telescope construction. It sent my imagination soaring! At the time, too, I thought that the Optical Shop would be around for a good long time, for already I wanted to return – and eventually did – to complete more and larger mirrors. To that end, I had the distinct good fortune to be employed at the Planetarium during the 1990's; so, I had the privilege to work on my own mirrors in the Optical Shop in my off-hours.

For nearly 40 years, the Optical Shop was a mainstay, glassed in like a living exhibit, offering visitors a great chance to watch amateur telescope-makers at their craft. Naturally, one might think such a fairly popular fixture would remain at the Adler Planetarium. Coincidentally, during the 1990's, the Planetarium underwent renovation and new construction; part of that renovation included dismantling the Optical Shop forever. At the time, the long-standing Chicago Astronomical Society stepped up and salvaged the shop, re-opening its setup later in an industrial park in Lansing, Illinois, where Seevers eventually ran its operation.

No doubt it was just a coincidence, but it was ironic to me that the Optical Shop, a once living exhibit that ran for 40 years, closed in the same year that the neighboring Field Museum adopted the same format and opened its own living exhibit, eventually known as the McDonald's Fossil Preparation Laboratory, which displayed museum technicians cleaning and assembling dinosaur bones. Closing the Optical Shop seemed to signal to the visiting public and a prospective ATM community that the Adler Planetarium was altogether finished with the art and science of telescope-making, that it was all done with mirrors. At the very least, that's how it appeared.

It's not a complaint that a happy haunt was eventually halted and sent packing. No, it's not that. In my experience, most ATMs mature from the experience of fashioning their first glass and seek more and greater challenges in telescope-making. Perhaps the Optical Shop – an exhibit of live persons working behind glass – was thought to be outmoded. If that was the case, then it might've been possible to re-purpose the Optical Shop into something less quaint and more robust, such as a modern mirror laboratory for testing commercial optics and astronomical products. Would the shop have attracted a dedicated following and more visitors to watch the action? We'll never know. More to the point, the Optical Shop likely was seen as prime real estate that, once reworked, would improve the flow of patrons through new exhibit space.

By the way, the Field Museum's fossil prep lab – an exhibit of live persons working behind glass – has been popular with the public ever since it opened.

* * *

What makes a Mersenne design work?

In mid-2004, I sought to redo the plywood prototype as a set of stowable, machined components that could be assembled in the field with a hex key or two. I figured that the design might benefit if its parts were reworked and re-purposed, so the need for the current prototype. Though several parts are indeed portable in their own carrying cases, none is particularly lightweight since I always intended to have a prototype prepared with an emphasis first on stability, and that meant heft. If that newer model presented unforeseen difficulties, I was not averse to correct its problems with usual construction techniques. So I moved ahead with that model anyway. By early 2006, that hefty prototype, worked on in my spare time, was complete. (See Photo 13, which shows a rough cut, fresh from the machine shop, of the current model.)

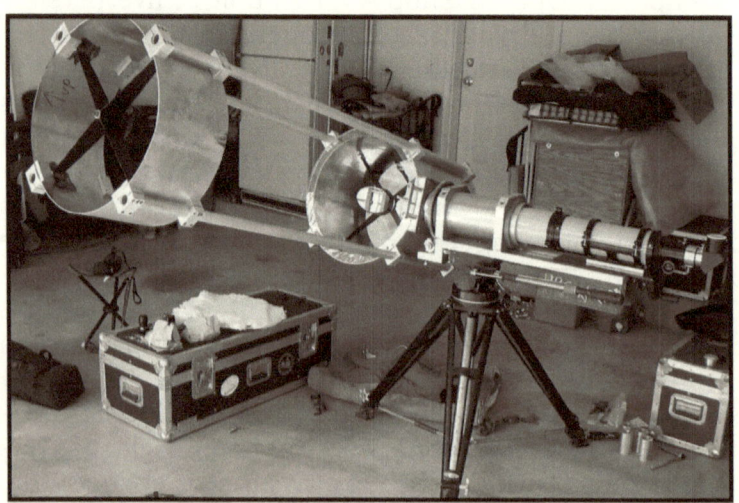

Photo 13

**This is the plywood prototype rebuilt as an aluminum prototype, almost finished.
Here the optics are set up for testing, circa 2004.**

During that time, I had machinists at a local shop fashion the parts that helped to complete the design. They milled and drilled a custom aluminum platform to receive the universal dovetail on which the refractor is mounted. They also prepared a hollow aluminum shaft as the main axle, a receiving flange that attaches the tube holding the optics to the main axle, and the large welded aluminum upper and lower tube assemblies to hold the optics. (See Photo 14, which shows the mount's substantial main axle.) In addition, four parallel thick-wall aluminum struts, firmly anchored in square compression slots, were fashioned to join the upper and lower tube assemblies, offering quick assembly and assuring rigidity whenever the telescope was set up.

Photo 14

Main axle close-up of the aluminum prototype.

The struts were slotted against inner stops to make certain they keep the mirror assemblies the exact same distance apart. The struts' original 70-inch lengths served as the correct distance to separate the mirrors by the requisite five feet for Mersenne-Nasmyth mode. Too, the struts were made with 38-inch extensions, slotted into round compression fittings (of aircraft aluminum), which provided the overall length to separate the mirrors by eight feet for Mersenne-Gregorian mode. Each strut was incised to match its compression fitting and straightened to a very fine tolerance so that the optical path remained true. (See Photo 15, which shows the lengthened struts set up for measurement at the local machine shop.)

Photo 15

The lengthened struts, needed for conversion to Gregorian mode, were straightened to a fine tolerance at a local machine shop.

When assembled, the main tube can move fairly easily to and fro by a solid array of steel and bronze gears accessible to the observer with the simple turn of a crankshaft. The upper tube assembly is counterbalanced as needed by a small rack of counterweights anchored beneath the main mirror. The entire tube assembly is itself counterbalanced by a set of adjustable inboard steel weights bolted beneath the main platform that holds the dovetail plate. The fully loaded tube assembly, the aluminum platform, its main bearing, the refractor, and counterweights total about 200 pounds when counterbalanced. (See Photo 16, showing full view of the current Mersenne-Nasmyth prototype.)

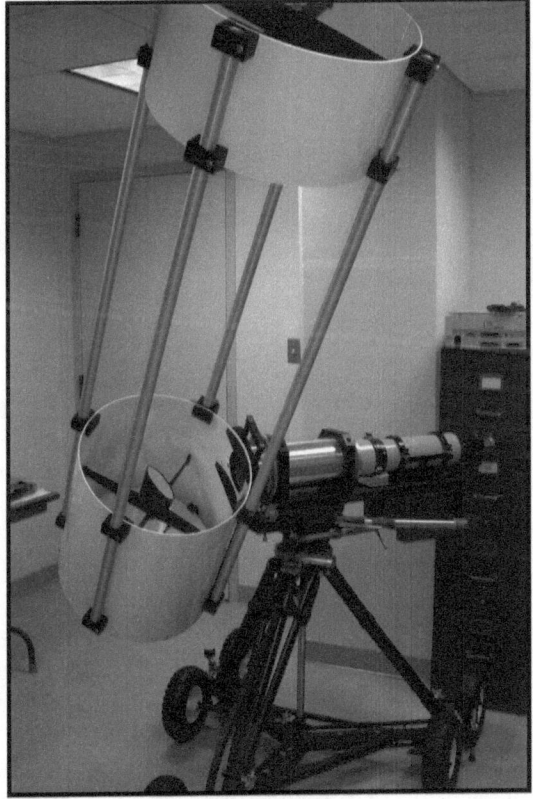

Photo 16

Full Mersenne-Nasmyth set up indoors.

The crankshaft also has gearing to accommodate a declination stepper motor with a clutch movement to disengage it when cranking manually in altitude. (See Photo 17, which shows a close-up of the stainless steel beveled gearing.) A surplus bronze gear and its hardened steel worm fitted to the tripod head accommodates the azimuth stepper motor used for right ascension. Both stepper motors slew and track just fine, with the scope correctly balanced.

Photo 17

Beveled gearing controlling the altitude crankshaft; round encoder attached in its own holder.

In 2008, a plate-and-clutch assembly for the azimuth motor was also machined to accommodate it on the tripod head. (See Photo 17 again, which shows the plate-and-clutch below the gearing in the lower right.) For a few finishing touches, all of the metal parts were buffed and polished. The tube sections were powder coated in white; the struts were anodized satin-clear while all other machined aluminum parts for support were anodized black. (See Photo 18, which shows an inside view of the lower tube assembly.)

Photo 18

Inside view of lower tube assembly.

As of this writing in 2011, these arrangements have made the Mersenne-Nasmyth design worthwhile to construct. Taken together, all of the metal parts have worked fairly well, making this particular version a good engineering model, a large step up from my plywood prototype *Styckehenge*. Out of this newer metal prototype, though, it indeed has become evident to me that this arrangement of optics could benefit from a further analysis of performance, transportability, strength of materials, and ease of set-up, which was the point of having this prototype built in the first place.

To test the Mersenne-Gregorian design as a working model, the convex secondary must be swapped with the concave secondary, the 38-inch extension struts must be inserted, and several small iron counterweights must be added to a rack beneath the main mirror. This latter arrangement could also benefit especially from the same analysis, due to the re-distribution of weight resulting from extending the struts when converting to Gregorian mode.

Already I anticipate that any upgrade or re-design of either operating mode would result in changes to four areas.

First, *performance in the field* can improve, especially if the design must re-distribute weight better to balance a swap of modes from Mersenne-Nasmyth to Mersenne-Gregorian. This would dictate relocating the center of gravity of the whole assembly closer to the ground, as in a typical Dobsonian design.

Second, *design strength* obliges a combination of lightweight materials, rather than entirely of metal. This might sound obvious; however, only when parts are tested working together does a clean assessment become evident.

Third, *ease of transport* suggests the ability to nest and store the telescope's components, just as was accomplished with *Styckehenge*. There are several practical ways to do this.

Fourth, *ease of set-up* requires that all optical components must align accurately with minimum fuss. Either mode of operation still calls for collimation of four mirrors; that is, the confocal, two-mirror arrangement and two additional flats. Though a total of four mirrors requires extra attention, their collimation is not unworkable. To facilitate this might require re-locating the position and angle of the receiving refractor at which it presently resides in the current prototype.

Taken together, the changes in these areas imply that a re-design might then resemble the arrangement which Clyde Bone himself used! While that may be possible, I believe it is more probable that any redesign, especially to make confocal mirrors useful in the field, will be less bulky than Mr. Bone's original telescopes. This is not a criticism of Bone's original scopes; rather, it's a natural improvement in the design's construction.

Now that I have used my current model, I have seen that its look and function can likely work well in a permanent location, say, for example, atop a solitary pier securely anchored within an observatory. Properly balanced, the current prototype has the heft for stable operation in that kind of setting. For a useful field model, though, the traditional look and basic function of the classic Dobsonian admittedly appears preferable. This is not to say I've circled back from where I started; rather, I have determined that the design of a hefty prototype on wheels lends itself fairly well to a literal roll-out into the field, but which is likely better suited for retirement to a permanent location, perhaps within a roll-off roof observatory.

Now one might ask: So, after all of this, a tale of two mirrors, you're starting over with the confocal design? No, more to the point, I continue with its advance. Prior to this, I've emphasized that telescope-making offers avenues of insight for imagination and innovation. I merely see that the Mersenne design is useful not only in more than one mode, but that it doesn't have to be built just one way. There's latitude for testing, for trialing, for tinkering. And so I simply see the next logical advancement, the next step, and a new target: a trimmer, lighter weight, more portable design.

Already I have plans afoot for that next step, which is the construction of a commercial prototype. To that end, let's see how the Mersenne configurations really work.

Chapter 7

The Victory Lap

Then comes that magic moment

When I finally zeroed in on the correct curve, I didn't know it at first. In figuring, that is, in crafting the parabolic surface, one must work in short stints, for the entire surface of the mirror is not necessarily worked at once. The mirror-maker might be looking to cure a turned edge or level a central peak. Correcting localized features requires skill, patience, and experience.

As my mirror neared the correct curve, I was simply following Ken Wolf's instruction, which usually consisted of a precise set of strokes and steps around the barrel before checking the mirror's curve once again. At the time, I was directed simply to make one more lap around the barrel – easy does it! – then clean and dry the glass, set up the mirror on the optical bench, let it settle to its surroundings, and come back in a while to look at its figure.

In hindsight, it was always surprising that he could very nearly name just the right stroke for the right amount of time and arrive at the right result. Of course that level of precision comes with first-hand know-how, but he didn't express that precision with a military exactitude. He didn't need to, as it was usually just a brief, easy-going instruction. But it was usually correct. When he looked up from the Foucault tester – he made the call rather plainly – it was clear that my mirror was finished. I was elated! I'd done it!

It was likely I wondered about my mirror's wave-rating – of course there's a way to calculate it – and he obliged with a fraction and I took his word for it. A wave-rating better than one-quarter – a minimal rating – is always sought, but I recall he said "one-tenth" which made me feel as if I'd really accomplished something. Years later, when I sought to have that little disk re-surfaced, Jim Seevers checked it with the shop's Foucault tester and confirmed both the smoothness of the curve and the validity of its wave-rating.

By the way, my mirror was finished on an auspicious day at the Adler Planetarium. Coincidentally, on the very day Ken Wolf called my mirror complete, a member of the Planetarium staff swept through the Optical Shop asking if ATMs working the barrels would like to come outside for an important event. Curious, we ATMs and a few other visiting patrons at the Planetarium that summer's day filed out to one of the big boxy granite slabs flanking the front entrance's grand staircase. The younger of us clambered atop. On a stretch of lawn in front of the granite slab, several gentlemen were standing, lined up with shovels, apparently waiting for their cue from a photographer. We sat behind them. The date was August 9, 1971.

The line of five gentlemen, neatly attired for business that Monday, comprised city officials and Planetarium dignitaries, including Robert Adler, son of the Planetarium's founder, Max Adler. Shovels in hand, they were ready for a groundbreaking, to turn over the first ceremonial clods of earth for the construction of the Astro-Science Center, the name at the time for a future addition to the Planetarium.

On that day when I completed my little mirror, it became evident the Adler Planetarium was moving ahead with big plans to expand its own universe with more exhibit space for the visiting public and the curious patrons sitting atop that boxy granite slab. If I'd expressly thought about it at that moment, it might be exciting to know how the Optical Shop might be part of that new Astro-Science Center. After the little ceremony – photos of which can be found online or in the Planetarium's archives – we curious patrons dispersed, perhaps each with his own idea of what the future may bring.

Once my first mirror was complete, it was plain to see what lay in its future. The rest of my telescope's construction fell fairly quickly into place. Of course the mirror was sent off to Clausing's for its coating in Beral, a hard, shiny alloy for front-surface mirrors. Yes, I did finish the telescope's mirror, but continued to work on a suitable mount. For that mount, I made use of several homebuilt kinds at first (including my construction of "The Tank") but when each kind proved problematic for one reason or another, the experience gave me insight toward the stability needed.

I settled on bronze-lapped pipe fittings prepared by a machinist and purchased two engraved, aluminum setting circles from Edmund Scientific. (See Photo 19 for a picture of my first 6-inch Newtonian after its completion in 1971.) This made the whole assembly fairly serviceable for years. But since I like to tinker, always trying to make it better, I rebuilt that first scope, giving it a better look, even greater stability, and a clock drive for celestial tracking.

Photo 19

**My homebuilt Newtonian on its
pipe fitting mount.**

Since that time of its first construction, I have sought to remain true to the intent of having a telescope that reminded me of the kind I'd always wanted when I completed that first mirror 40 years ago. So, years ago when I opted for an upgrade of the whole instrument, I stayed with materials of the day, circa 1971, with which to re-do the telescope; subsequently, I procured two valuable finds, namely, a classic equatorial mount from Kenneth Novak, a well-known purveyor of telescope parts at the time, and a brand-new Byers clock drive, the very same kind I'd once seen advertised on the pages of *Sky and Telescope*. (See Photo 20 for a picture of my first 6-inch Newtonian, rebuilt.)

Photo 20

My homebuilt Newtonian on its Novak mount, 2011, but with parts circa 1971.

This whole description may sound unnecessarily detailed, but, like an automobile aficionado restoring a classic car, it was certainly worth the effort to spend the time to get this accurate and looking good. Though I've built a modest number of telescopes in my day, and have had one happy time building them all, I still regard this first scope, my 6-inch Newtonian, as one of my more memorable endeavors in telescope-making.

* * *

How does this Mersenne design really work?

In and of itself, the optical configurations suggested by Marin Mersenne appear close to pure genius. That is, what makes the Mersenne-Nasmyth or the Mersenne-Gregorian memorable is the distinctive feature the observer sees on peering into the eyepiece. Simply put, confocal mirrors bundle incoming light to a narrow path that contributes magnification without producing an image. While that may not sound significant or even useful, a confocal mirror system becomes an effective light-beam compressor. And it is that exceptional feature of Marin Mersenne's design that makes it so appealing and promising in telescope construction.

From the details on this design available from the optics literature, evidently the ideal Mersenne design is *aplanatic*, so it does not suffer from spherical aberration, which means that one can observe a clear, undistorted image. A Mersenne design is also *anastigmatic*, so it does not appear to suffer from astigmatism, which is an optical defect that prevents light rays from meeting at a single point. Finally, a Mersenne design is *achromatic*, so it bends light without dispersing it into its component colors, which helps to deliver images that lack extraneous artifacts of color. When coupled with a well-made refractor – preferably one that is free of color defects – positioned to receive the compressed light beam, the result greatly amplifies the virtual aperture of the refractor, making a modest refracting telescope behave as if it has the wider aperture and greater light-gathering power of a large reflecting telescope.

For the serious amateur there is sufficient evidence for the feasibility of either a Mersenne-Nasmyth or a Mersenne-Gregorian. Mr. Bone notes it in his *Sky and Telescope* article, and I reference it here. In Edward Hubert Linfoot's book, *Recent Advances in Optics*, the author details numerous combinations of multiple-mirror designs. In Chapter 4 of his 1955 edition, he specifies two-mirror systems in particular and provides a dense mathematical treatment for the existence of two-mirror anastigmats with a concave primary. He expresses mathematically the conditions needed to satisfy the equations that describe these optical systems. His conclusion is that such systems with a large primary are possible when they consist of two confocal paraboloids (either convex or convex secondary) and forming an image at infinity, which describes ideally and optically the Mersenne design.

And, as mentioned earlier, the recent works by Wilson (in 2007) and Lemaitre (in 2009) confidently attest to the untried worth of either the afocal Cassegrain or Gregory design. Even though either configuration is aplanatic, anastigmatic, and achromatic, these authors conclude that these afocal systems are rarely appreciated. A modern unfamiliarity with past optical designs appears to have precluded their use.

Moreover, they suggest that Mersenne and his contemporaries did not – and could not – realize how advanced were these afocal configurations. That is, Mersenne's designs would eventually show optical principles that were unknown at the time to Mersenne and his associates. With either his confocal Cassegrain or confocal Gregory, Mersenne proposed designs that very nearly obviated the optical defects that later opticians would seek to minimize. More to the point, both authors express irony and surprise that Mersenne's designs escaped the notice of the best known of the modern mirror-makers.

There's one other quick item to address. In my current prototype, since the light path in a Mersenne-Nasmyth design is directed to a Nasmyth focus for magnification, rather than passing through a perforation in the primary as in a conventional Cassegrainian or Gregorian, the observer remains nearly five feet from the main optics. At that distance, the observer does not get in the way of the tube nor does one's body heat interfere with the light path.

One might think that last item would not influence much at all; rather, nearby heat from a human body or residual heat from a telescope equilibrating to the night air can play havoc with the light path. The image at the eyepiece is likely to show more scintillation than one might find on a clear night with bare eyes alone. Excess or residual heat has been such a concern in mirror-making that Ingalls included an essay in ATM Book One (pages 147-148) on the use of insulating materials to keep heat away from the telescope's tube and eyepiece.

Now, let's look at the scope's operation. As of this writing, the current prototype offers a sturdy base. The main platform, on which the entire tube assembly and refractor balances, is a well-crafted design, offering a fairly stable platform that allows easy maneuvering. With this present arrangement, it is likely that an observer's well-made, short-focus commercial refractor, up to 102 mm (i.e., 4 inches) aperture, can be accommodated on the platform's universal dove-tail. For apertures greater than 94 millimeters – the refractor's aperture should not exceed the minor axis of the elliptical tertiary – the scope's dewcap should be removed so the scope's main tube itself can slide cleanly into the mount's main axle. And to complete the arrangement, occasionally a Celestron C90 mounted skyward on the lower tube assembly serves as a finderscope.

Location of night-sky objects is relatively straightforward as with any altazimuth mount, but is aided with the addition of a Sky Vector system and encoders. One encoder is anchored in a neat, trim holder beneath the azimuth bearing of the tripod to read right ascension; the other is slotted in its own holder to read declination. The latter encoder is reprogrammed to read properly at a ratio of four-to-one since it is mounted onto the crankshaft that turns a worm to move the geared main axle. (Go back for a moment to see Photo 17.)

One curious effect of using confocal components is that the refractor receives two images, a "high-power" view superimposed on a "low-power" view for the same eyepiece. This sounds disconcerting but can be put to good use, namely, by initially using that "low-power" view as one's finderscope image and then observing the actual "high-power" image. By the way, I know the more correct term is *magnification*, rather than the word *power*; but, here I use the latter word in its everyday sense, and I don't think you'll be confused by it.

In the following set of two sample photographs of the Moon, note the effect of superimposition. In the first photo, that is, the "low-power" view of the waning gibbous Moon, one sees the entire Moon as the refractor views the sky only with the aid of the tertiary mirror, which is a flat diagonal. (See Photo 21.) But rack in the refractor's drawtube with its focuser and the same eyepiece then intercepts the compressed light beam from the primary and secondary mirrors, presenting a "high-power" view of the same Moon. The next lunar photograph depicts this magnified effect, showing the central portion of the same waning gibbous Moon along its terminator. (See Photo 22 for the magnified lunar image.)

Photo 21

The entire waning gibbous Moon taken through refractor on Mersenne-Nasmyth telescope.

Photo 22

Waning gibbous Moon, mid-latitudes, taken with Mersenne-Nasmyth telescope, same time, same eyepiece, but viewed with confocal mirrors.

To demonstrate this effect, the previous two photos are unretouched images taken afocally with a digital camera. More to the point, the two photos are separate lunar images taken at nearly the same time; that is, the latter one is not an enlargement of the first photo. That latter photo was taken mere seconds apart – no changes – and then cropped to the same image size for this chapter. By the way, these two photos were taken through the same Kellner eyepiece of 18 millimeters focal length.

To see this effect more clearly, note the next set of photos, shot in the daytime. To set the scene, the first of these photos shows the rooftop of a nearby house. (See Photo 23 and note the rooftop flue in the center.) The next two photos were taken afocally (i.e., with the same digital camera and through a Plössl eyepiece of focal length 20 millimeters), which show close-ups of the tall flue. The first photo of the pair shows the flue as seen through the refractor by the tertiary, that is, the "low-power" view. The second photo, taken with the same camera, shows the same flue through the same eyepiece, as seen by the confocal mirrors, that is, the "high-power" view. (See Photos 24 and 25 for "low-power" and "high-power" views of rooftop flue.)

Photo 23

Rooftop flue of neighbor's house. The taller one is our subject.

Photo 24

Rooftop flue as seen in "low-power."

Photo 25

Same flue as seen in "high-power."
Same eyepiece used in previous photo.

The pair of photos shows one reason why an unbaffled Mersenne is not well suited for daytime observation. In the "low-power" view, one essentially looks through the "high-power" view, not in focus. This makes the "low-power" view appear slightly foggy because the field of view has two images, one of which is unfocused and superimposed on the other. By the way, for this photo I did defocus to a point at which to enhance the effect of superimposition. Once the secondary mirror is baffled, though, the superimposition of the "low-power" view disappears and then one can only view in "high-power."

The unbaffled "low-power" Mersenne view is expected, and is easily explained, as it shows what one normally sees at modest magnification through a well-made refractor. But, to a casual observer, the "high-power" view is completely unexpected. Yes, it is true that a Mersenne configuration of confocal optics delivers this hugely enlarged image to the eyes of the observer – that's a distinct advantage – but one really has to see it to believe it.

This "high-power" view is far larger than any image possible with the same refractor and the same eyepiece. Indeed, the confocal optics gives the relatively small refractor's aperture a greater virtual light-gathering power that it normally does not possess. And this is due to the much larger aperture of the mirror of the confocal, two-mirror system helping to deliver a magnified, collimated light path. One can clearly see that marrying a good refractor to the large aperture of confocal mirrors makes that refractor behave as if it has the aperture of the large mirror!

Of course, this optical feature of superimposition can be confusing at first. In effect, the "high-power" view merges with the "low-power" view. Yes, baffling can eliminate this confusing feature. But, when viewing the night sky with an unbaffled Mersenne, one doesn't necessarily see the merge of one image superimposed on the other. Only when looking at a bright object, such as the Moon, does a superimposed image come with unwanted glare; again, one can eliminate optical artifacts with appropriate baffling of optics. If you're an avid lunar observer, baffling the secondary is necessary.

A simple approach to baffling involves the crafting of a shallow, thin collar for the circular rim of the secondary mirror. Mr. Bone himself describes his own procedure where, by trial-and-error, a series of cardboard baffles, increasing in size by half-inch increments, can be made and tested to eliminate the off-axis light rays that contribute to the "low-power" view.

Baffling the tertiary is a good idea, too, but with a simple sleeve (not a collar) extending beyond the mirror's major axis for no more than an inch, to block off-axis light rays. Ray tracing for precise baffling can help to accomplish this, or, one can consult an online baffle calculator. Several prospects are listed at the end of this little book.

So far, I've found that certain eyepieces, viz., Plössl, Erfle, Orthoscopic, and Kellner, work well enough to deliver the appearance of a relatively flat field to the observer's eye. That is, with the use of these types of eyepieces, there appear no bulges and few blooms of light in images seen in the field of view. Precise ray tracing can confirm this, though apparently, as of this writing, nowhere in the popular literature has this task been done analytically, which is why this aspect of any Mersenne design is still trial-and-error. A future project may involve testing more eyepiece designs and their name brands.

While a few unknowns in overall performance need further examination, I have no doubt that, though the Mersenne optical configuration is workable, the operation of my current prototype will need to be upgraded, as I'll always seek to improve it. Its function is sound; though, as mentioned in the previous chapter, the optical configuration and use of materials can always be improved and will likely be re-purposed for better use in a different setting.

In the meantime, don't let the idiosyncrasies of either of Mersenne's two-mirror designs thwart your own attempts to craft a workable telescope. Though his designs may not be exactly what you have in mind, they certainly can show the serious amateur what one can attempt on one's own. Years ago, the Mersenne telescope seemed like it wouldn't happen, as I toyed with glass, wood, and metal. Occasional skeptics leaned in to take a look at the design and offered their criticisms but not their encouragement. The design seemed too big, too cumbersome, too complicated, and that might've stopped an ATM from going any farther.

It is my experience that too often the mirror-maker can think things are too hard and stops short of completing his goal, improving his design, or finding the right scope for what he really wants to observe. Skeptics shall always remain skeptical; critics will always criticize. Ignore them. Build your telescope, whether it's my big, cumbersome, complicated design or your own homebuilt version, just build it. Do the best you can with it. Be proud of it. Then, if you must, take your victory lap.

Chapter 8

One Happy Time

Gilded memories

Every amateur telescope-maker doubtless has a time in his life when memories are made that last a lifetime. My turn around the barrel at the Optical Shop of the Adler Planetarium was indeed a memorable time for me and for untold others who joined in the fun. But the fun didn't end when the Optical Shop closed. From the mirrors that I made, from the telescopes that I built, came the nightly enjoyment of using those very instruments, crafted by my own hand, with which to observe the night sky. It is a satisfying sense of accomplishment that accompanies any hobbyist's achievement of success in his avocation.

For those who seek their own memories of a rewarding time with the night sky, the memories are even more gilt-edged when you make a great effort to achieve your goal of a hand-crafted instrument. Not every hobbyist may seek to go this far, but your own efforts are well rewarded in every sense when you design your own device, make your mirror, assemble your own components, and share your results with those around you.

In the realm of amateur telescope-makers, one doesn't need to brag to share. Your own handiwork, made to the best of your ability or with what materials will allow, speaks to your dedication to your hobby. A welcome remark or a knowing nod from a fellow enthusiast lets you know your instrument has boldly, tacitly, made your point for you.

* * *

Make new memories for yourself

For the most part, I am pleased with the current instrument, as it serves well enough for nightly sessions under the stars. But there is always the temptation to tinker, to make it better. Not that I don't leave well enough alone; rather, after constructing this instrument, a prototype after all, I discover more to do with the design.

Overall, this current arrangement appears adequate but wants for two things, ease of collimation and even greater stability. Having a total of four mirrors requires precise collimation. Such a task is not impossible in the field – it just can be a little tricky, and it can take a fair amount of time to get it exact. A laser collimator is a good start toward precise collimation.

To align optics on merely a mechanical level does not itself consume much time, and ought to be done in daylight. I start with the refractor removed from the main axle, so that I can see the primary and secondary reflected in the tertiary. Collimation of a four-mirror configuration starts with the primary, where the thin vanes supporting both the secondary and tertiary mirrors must appear in line, assuming, of course, these two mirrors are already centered mechanically within the tube assembly. Since the current prototype has an upper and lower tube assembly, the rigid struts, straightened to their fine tolerance, help to keep both aligned. Adjusting the screws or knobs behind the primary's holder will align the vanes. Finer adjustments to the secondary and tertiary will center the remaining reflections. (See Photo 26, for a view of fairly collimated optics.) Then I insert the refractor to complete the process.

Photo 26

Daytime view through the main axle, refractor removed, showing nominal alignment of the mirror components as they near collimation.

My refractor's quartz diagonal, which is the design's quaternary mirror, then can be viewed to center the target of reflections of the three other mirrors. After nightfall, the laser collimator comes in handy, using the typical pattern projected by the collimator placed in the eyepiece position. The laser pattern, viz., a grid, is then centered on the tertiary, but care must be taken to adjust the tertiary in tilt, too, to center correctly the pattern, not only on the tertiary, but onto the secondary. The laser grid, once centered on the secondary, should then be adjusted to center the pattern on the primary. Admittedly, this procedure likely will need to be repeated to get collimation correct. Also, the laser pattern, a grid that spreads out with distance, may be faint to see but should work with a little practice. Repeat as necessary for each observing session.

As both scope and mount are level but weighty, hanging the entire optical tube assembly on one side of a stout tripod's azimuth bearings requires adequate balancing on the other side. The counterweights on that other side are inboard and can slide to adjust to the load, but, once balanced, the metal assembly on its bearings can acquire its own moment when unlocked for use. And the scope has its own frequency; it quivers. Though this is not unexpected for a metal assembly swinging on a wheeled tripod, it can be addressed with a different arrangement of components to make the whole assembly less lively.

For example, if anchored in an observatory setting, I expect the liveliness can be dampened with rubberized pads beneath the main platform, remounted atop a concrete-filled steel pier. Nothing surprising there. But for better field use, as outlined previously, my alternative for stability would be a large lazy-Susan altazimuth, much like a rocker box for a Dobsonian, only beefier as needed for a receiving refractor, as tall as needed for seated height.

As a telescope built for comfort, mine is a good optical instrument for the time being, mobile on wheels and portable once disassembled into its component pieces and reassembled in the field. Though it is chiefly operated in Mersenne-Newtonian mode, it can be fairly readily converted to Mersenne-Gregorian mode as part of its hybrid design. If it were to become a telescope built for the prospect of research-grade interest, more work is definitely needed but that work holds no surprises.

I believe that the telescope can be a hit at local gatherings, since parents no longer need to lift their children to the telescope's eyepiece, no one needs to climb a ladder to look through the eyepiece, and, those in wheelchairs can have an easy time when gazing at the night sky through a telescope at just the right height. Maybe you'd like to have that same opportunity to view comfortably. If so, spread the word that a feasible design already exists for you and others to enjoy the night sky in relaxed ease. Comfortable viewing certainly leads to more memorable nights and many good times, many happy times, under the stars. Start now to make those memories happen.

* * *

As of this writing, the Mersenne-Nasmyth design and its variant, the Mersenne-Gregorian, are being examined for their commercial viability. For now, I simply wanted to get out the word on a useful, workable, distinctive design, hence the aim of this little book. And there's good reason to do that.

In time, I anticipate that either configuration, the Mersenne-Nasmyth or the Mersenne-Gregorian, has a decent future with serious amateurs based on the two following observations. First, modern Dobsonian telescopes can likely be retrofitted to operate, for example, as a Mersenne-Nasmyth by replacing the existing Newtonian diagonal with a confocal convex secondary mirror, by inserting tertiary and quaternary mirrors, and mounting a receiving refractor to the existing rocker box. This first observation sounds fairly involved, but it is feasible. Second, the manufacture of a true Mersenne-Nasmyth, not necessarily a retrofit, has the potential of a serious following as it's always ready for use at just the right height for most observers. While the latter observation may sound obvious, it's not a trivial notion, especially if one doesn't have usual access to a telescope's eyepiece.

Anticipating that the uncommon Mersenne design can become popular may sound like wishful thinking, but I have good grounds to expect it can come to pass. Because I, myself, have already seen it. And, as far as one can tell, so has Clyde Bone, one of the first modern makers of this design.

For instance, my original Mersenne-Nasmyth, *Styckehenge*, was itself a retrofit. That telescope was a Newtonian at first, just like Mr. Bone's own scope. One could swap out the Newtonian diagonal with a convex confocal secondary and adjust its distance from the primary to convert it to Mersenne-Nasmyth mode. The tertiary and quaternary mirrors were already incorporated into the design with the receiving refractor. That is also how Mr. Bone's first Mersenne scope worked as well. As of this writing, I conclude it is likely that certain, selected, commercial Dobsonian models can be candidates for retrofit with an appropriate commercial kit.

Moreover, interested observers have seen my own current prototype as a boon, since, as just stressed above, parents no longer need to lift their children to the telescope's eyepiece, no one needs to climb a ladder to look through the eyepiece, and, those in wheelchairs can have an easy time when gazing at the night sky through a telescope at just the right height. Allowing for observation while seated was the same reason Mr. Bone offered for constructing his own Mersenne scopes at a comfortable height.

Commercial designs do not need to be wedded to certain formulas of construction. Though templates are often needed to aid quick and consistent fabrication of designs, it is very likely that the knowledgeable ATM can come up with a distinctive blueprint that offers stability and portability, ease of set-up and ease of access, comfortable seating and no ladder, functional operation and first-rate looks. Come up with your own good design, have a good time with it, and roll it out for others to try.

With my own first Newtonian, I had the good fortune to have good guidance. What good guidance gave me was the insight to tinker, the curiosity to explore, the need to improve, and the patience to find my own way. What I take away most from the crafting of that first mirror, that first telescope, are not just gauzy reminiscences of people and places; rather, it is one more set of experiences that offered insight, sparked imagination, and honed practicality.

With my own Mersenne-Nasmyth and its optical complement, the Mersenne-Gregorian, I have had the good fortune to see those designs reward me, at the very least, with a convincing sense of genuine accomplishment. Perhaps that accomplishment can offer you the same rewarding experience, if you seek to try either design for yourself. (See Photo 27, for someone pleased with his own design.) Good luck and let the good times roll!

Photo 27

**The author with his Mersenne telescope.
Photo by Denise Hurley.**

Epilogue

By 2011, nearly a full 55 years after its fortuitous inception as a full-fledged facility, and fully 40 years after completing my own first mirror, the Amateur Telescope Makers Optical Shop continued to inspire and instruct both novice and veteran telescope-makers in the art of mirror-making. Released years ago from the auspices of the renamed Adler Planetarium and Astronomy Museum, and since its actual move out of the Planetarium, the Optical Shop operated under the general sponsorship of the Chicago Astronomical Society.

As he directed the Optical Shop when it was housed at the Adler Planetarium, Jim Seevers continued overseeing its activities with unstinting commitment. Till recently, and from its unassuming digs in Lansing, Illinois, the Optical Shop sought to carry on an extraordinary tradition begun by Russell Porter in Springfield, Vermont, introducing amateur astronomers to their own future reminiscences in the art of mirror-making. As of this writing, according to Seevers, the tradition will continue, but at a new location and under new supervision. It is good to know the Optical Shop lives on.

It's good to know, because, on occasion, I've read and heard claims that amateur telescope-making reached its zenith many years ago. I disagree, and here's why. Though many of us may have fond first memories of mirror-making from decades past, the avocation hasn't remained frozen in time. Amateur telescope-making is not just the purview of the gentleman scientist or a throwback to an earlier era. It is still very much modern. It has advanced as we ATMs have matured. And as we have grown up with our hobby, others have come along and joined with us to share both the fascination and satisfaction of crafting their own instruments with which to view the heavens.

It is from the ranks of amateur glass pushers, young and old, skilled or novice, that newer shop techniques or construction ideas or optical designs or imaging techniques or methods of adaptive optics have emerged to become the new standards. And the practitioners of those standards have often shown themselves to be the next generation of entrepreneurs that have helped to advance the art and science of crafting first-quality telescopes.

From the perspective of my own generation, my mirror-making days have been most satisfactory. At those times when I thought it best, I built the telescopes I wanted for the reasons I needed them. And when I got good results, I had the desire to pursue even better results, bigger mirrors, and more challenging telescopes.

Moreover, that desire for ever larger aperture and its expected reward did indeed inspire a meandering search for the ideal telescope. Though that search narrows every year, its desire does not. Yes, it's unlikely to find that true ideal, but that ultimately does not become the endpoint of the search, for there can always be a fresh target, a new standard, to pursue. More to the point, it's the adventure of getting there, and finding what can be attained, that is more than adequate to satisfy one's desire and need for the moment.

My own search for a workable telescope gave way to something homebuilt but far better than a commercial product I would've wanted to buy. My own search found me probing several directions until one seemed suitable. My own search resulted in a confident sense of success in the instrument I eventually crafted. My own search in the long run aimed me in a direction that allowed me to craft something original and to my own satisfaction. My own search led me to a new ideal, which continues to be improved. That's just part of the adventure.

And so the adventure goes on. And that's just the way many of us like it.

References

Notes, Books, Magazines

Recent Advances in Optics, Edward Hubert Linfoot, Oxford University Press, London, 1955.

> This book, also referenced by Clyde Bone in his *Sky and Telescope* article (listed below), presents the most complete mathematical treatment of the Mersenne design. Though the design is not invoked by name, it is described generically in Chapter 4, page 274, under Two-Mirror Systems. This book is out of print but can be obtained through inter-library loan. It is not necessary to have this book to complete the Mersenne design, though it does have useful mathematical insights for determining not only the distances between components in two-mirror systems but the formation of real and virtual images by these systems.

Making Your Own Telescope, Allyn Thompson, Sky Publishing Corporation, 10th printing, 1973.

> Of course there is no mention of either Mersenne variant in this classic tome, first

published in 1947, but Thompson does offer complete information for making a telescope virtually from scratch. I include it here since I mentioned this book in the foregoing text.

Amateur Telescope Making, Book One, Albert Ingalls (ed.), Scientific American, Inc., New York, 4th edition, 24th printing, Kingsport Press, Inc., 1980.

The Mersenne designs (either Mersenne-Nasmyth or Mersenne-Gregorian) are not considered in any of the three ATM books of this series, but the discussion that starts on page 215 (Part X, Chapter 1, Compound Telescopes) in Book One is useful for comprehending the optics of a Gregorian system.

How to Make a Telescope, Jean Texereau, 2nd edition, Willman-Bell, Inc., Richmond, VA, 1984.

The Nasmyth design is addressed in Chapter 6, starting on page 142, but with an eye toward its use in a Cassegrainian configuration. Texereau does point out the Nasmyth feature is used on large telescopes in professional observatories. He makes no mention of its possible connection as a variant linked to the Mersenne optical concept of confocal mirrors. In Chapter 12, starting on page 237 ("Designs to be Avoided"), it is heartening that Texereau does not list a Mersenne variant as an optical configuration the ATM should not pursue!

Telescope Optics, Harrie Rutten and Martin van Venrooij, Willman-Bell, Inc., Richmond, VA, 5th printing, 2002, p. 140.

The authors do present a brief examination of the Mersenne design, starting on page 140, and conclude that the parallel beam produced by confocal mirrors can likely be funneled to another optical device. This is precisely both the intent and use of Mersenne designs, and it is the reason I find Mersenne designs so appealing. In an ensuing mention of Gregorian designs, on page 143, the authors briefly point out that the classic Gregorian has the prospect to produce an aberration-free visual telescope by matching the field curvatures of the Gregorian to the selected eyepiece; however, this is likely to involve trail-and-error for a good match. No mention is made of Mersenne variants that involve the Nasmyth focus or a Gregorian optical configuration.

Reflecting Telescope Optics: Basic Design Theory and Its Historical Development, Ray N. Wilson, Springer-Verlag Berlin, Heidelberg, 2nd ed., 2007, pp. 4-6, ISBN 978-3-540-40106-7

This is a very good recent source of detailed and summary information on various popular and lesser known reflective telescopic systems. The treatment of optical systems is heavily mathematical, though the individual descriptions are relatively straightforward and well cross-referenced. Here, one learns of the distinct optical advantages of the Mersenne system and its variants as first presented by Mersenne. Portions posted online through Google Books.

Astronomical Optics and Elasticity Theory: Active Optics Methods, Gérard René Lemaitre, Springer-Verlag Berlin, Heidelberg, Corrected 2nd printing, 2009, p. 14, ISBN 978-3-540-68904-1

> This is also a very good recent source of detailed and summary information on various popular and lesser known reflective telescopic systems. The treatment of optical systems is heavily mathematical, though the individual descriptions of optical systems are relatively straightforward and well cross-referenced. Here one learns that Mersenne was well ahead of his time, but did not recognize entirely the distinct properties of his afocal two-mirror system. Portions posted online through Google Books.

James Nasmyth, Engineer, An Autobiography, James Nasmyth, Samuel Stiles (ed.), John Murray (pub.), London, Printed by R & R Clark, Edinburgh, 1883 (digitized by Google Books), pp. 172-198.

> The author provides an assured, chatty, detailed account of his life and times, his family's ancestry, and historical context of his inventions and interests.

"A Novel Dual-Field 8-inch Telescope," Wesley N. Lindsay, *Sky and Telescope Magazine*, Gleanings for ATMs, February 1965, pp. 112.

> The first known modern description of the Mersenne design, prepared by an ATM, appears here.

"Twenty Inch Genesis Handiscope," Clyde Bone, *Amateur Astronomy Magazine* (#4), Tom Clark (ed.), Winter 1994, pp. 45-47.

> The Mersenne-Nasmyth design is described herein as a telescope combined with a standard Newtonian as a Dobsonian configuration. That is, the scope can be converted from one design to another by swapping secondary mirrors and adjusting the focal length within the upper cage assembly. The scope's components disassemble and nest within the oversized rocker box.

"The Mersenne Telescope," Clyde Bone, *Sky and Telescope Magazine*, September 1999, pp. 130-133.

> The Mersenne-Nasmyth design is again presented by Bone, but with a larger version of his original combination. It's worth reading this short article as it summarizes most of the important aspects of construction.

"A Large Telescope of Dual Perspective," Roy Kaelin, *Amateur Astronomy Magazine* (#35), Tom Clark (ed.), Fall 2002, pp. 42-44.

> In this article, I present my own plywood version of Bone's combination telescope and the prospect of improving the design.

"Take a Seat and Enjoy the View," Roy Kaelin, *Amateur Astronomy Magazine* (#39), Tom Clark (ed.), Fall 2003, pp. 54-56.

> In this article, I elaborate on the good points and bad points of the Mersenne design, its

construction, and the usefulness of a Gregorian version of this design.

"The Comfortable Catadioptric," Roy Kaelin, *Proceedings of 46th Annual GLPA Conference*, South Bend, IN, on deployment of the original Mersenne-Nasmyth telescope, October 2010.

To my colleagues in the planetarium community, I present an overview of the Mersenne-Nasmyth configuration, its background, and the prospect of its use as a valuable addition to star parties and classrooms.

"A Fresh Look at a Catadioptric Design," Roy Kaelin, *Amateur Astronomy Magazine* (#69), Charlie Warren (ed.), Spring 2011, pp. 51-55.

In this article on the current aluminum prototype, I provide an update of both the Mersenne-Nasmyth and the Mersenne-Gregorian designs, their convertibility from one operating mode to another, and plans for improving their function.

Online Links for Further Reference

(As of this writing, these listings do not constitute product endorsements.)

Sources of optical ray-tracing and telescope design software:

http://optics-lab.com/

http://www.math.ucsd.edu/~sbuss/MathCG/RayTrace/

http://www.willbell.com/tm/tm6.htm

http://lambdares.com/education/

http://bhs.broo.k12.wv.us/homepage/alumni/dstevick/software.htm

http://www.astronomy.net/articles/31/

http://www.myoptics.at/jodas/twomirror.html

Sources for constructing baffles:

http://mysite.verizon.net/rflrs/nbaffle.htm

http://mysite.verizon.net/rflrs/baffle.htm

http://www.willbell.com/tm/tm6.htm

http://astrotips.com/Article71.phtml

Archive photo of Adler Planetarium, groundbreaking for Astro-Science Center, August 9, 1971:

http://www.flickr.com/photos/adlerplanetarium/4420734610/

Archive photo Adler Planetarium, a scene in the Optical Shop, noting its run at Adler Planetarium:

http://www.flickr.com/photos/adlerplanetarium/4420734278/in/set-72157623587570094

Threads of past discussions on Mersenne designs:

http://www.atmlist.net/ARCHIVES-1995-2004/1996/NOV96/msg00626.html

http://www.atmlist.net/ARCHIVES-1995-2004/1996/NOV96/msg00600.html

http://www.atmlist.net/ARCHIVES-1995-2004/1996/NOV96/msg00624.html

http://www.atmlist.net/ARCHIVES-1995-2004/1996/NOV96/msg00594.html

http://www.atmlist.net/ARCHIVES-1995-2004/1996/NOV96/msg00619.html

http://www.quadibloc.com/science/opt0203.htm

http://www.cloudynights.com/ubbthreads/showflat.php/Cat/1,2,3,4,5,8/Number/3881929/page/95/view/collapsed/sb/9/o/all/fpart/1

http://www.cloudynights.com/ubbthreads/showflat.php/Cat/1,2,3,4,5,8/Number/3952850/Main/3913270

www.ingramcontent.com/pod-product-compliance
Lightning Source LLC
Chambersburg PA
CBHW030843180526
45163CB00004B/1434